A GUIDE TO THE STUDY OF **soil ecology**

contours: studies of the environment

Series Editor
William A. Andrews
Associate Professor of Science Education
The Faculty of Education
University of Toronto

A Guide to the Study of ENVIRONMENTAL POLLUTION
A Guide to the Study of FRESHWATER ECOLOGY
A Guide to the Study of SOIL ECOLOGY
A Guide to the Study of TERRESTRIAL ECOLOGY

A GUIDE TO THE STUDY OF

soil ecology

Contributing Authors:
Nancy D. Davies
Daniel G. Stoker
Douglas E. Windsor
M. Terry Ashcroft
M. Carolynn Coburn
William A. Andrews

Editor:
William A. Andrews

Prentice-Hall, Inc., Englewood Cliffs, New Jersey

Library of Congress Cataloging in Publication Data

ANDREWS, WILLIAM A.   1930-
    A guide to the study of soil ecology.

    (Contours: studies of the environment)
    Includes bibliographies.
    1. Soil ecology.   I. Davies, Nancy D   1945-
II. Title.
QH541.5.S6A5   574.5'26   72-11522
ISBN   0-13-370973-6
ISBN   0-13-370965-5 (pbk)

A Guide to the Study of
SOIL ECOLOGY
© 1973 by W. A. Andrews
Published by Prentice-Hall, Inc.
Englewood Cliffs, New Jersey.
Printed in the United States of America.
10 9 8 7 6 5 4 3 2 1

Prentice-Hall International, Inc., *London*
Prentice-Hall of Australia, Pty., Ltd., *Sydney*
Prentice-Hall of Canada, Ltd., *Toronto*
Prentice- Hall of India Private Ltd., *New Delhi*
Prentice-Hall of Japan, Inc., *Tokyo*

Design by Jerrold J. Stefl, cover illustration by Tom Daly,
text illustrations by James Loates.

Title page photo reproduced from U.S. Department of the Interior Conservation Yearbook #5.

PREFACE

This book has two basic purposes. The first is to help you explore a world that may be strange to you—the world of soil organisms. Although you walk on soil just about every day, you may be familiar with only a few of its inhabitants. You probably know a fair amount about earthworms and soil-dwelling mammals like gophers. But you probably know little about the many hundreds of other fascinating creatures that live in the soil. Because they cannot move very quickly through the soil, most of these animals are well camouflaged. Special techniques are often required to isolate them from the soil. You will learn how to capture, identify, and culture many of these soil organisms.

    The second purpose of this book is to help you to learn the basic principles of ecology. Ecology is the study of the relationships between organisms and their environments. We need not impress upon you the need to study this subject. Just about everyone knows that, ultimately, survival of the human species depends upon all of us understanding and obeying the basic principles of ecology. You can learn these principles by studying a stream, a pond, a forest, or a meadow. You can also learn them by studying how the organisms in the soil affect one another and how they interact with their environments. This book will guide you as you explore these relationships.

ACKNOWLEDGMENTS

This program was developed at the Faculty of Education, University of Toronto. The resources of the Faculty and the knowledge and skills of many student-teachers in the Environmental Studies option contributed greatly to the quality of the materials contained in this book and its companion volumes.

The authors are particularly appreciative of the competent professional help received from the Publisher. In particular, we wish to acknowledge the editorial assistance of Sue Barnes, Carolyn Tanner, and Jane Standen. Their skill, knowledge, and patience are greatly appreciated. To Paul Hunt, Kelvin Kean, and John Perigoe we extend our sincere thanks for their help in planning this program in Environmental Studies. We are also grateful to Ron Decent and the many other members of the Prentice-Hall production staff for their effective work in the production of this book.

We wish also to thank Jim Loates for his excellent art work and Lois Andrews for her careful preparation of the manuscript.

N.D.D.
D.G.S.
D.E.W.
M.T.A.
M.C.C.
W.A.A.

CONTENTS

FIELD AND LABORATORY STUDIES   98

5

RESEARCH TOPICS   168

6

CASE STUDIES   180

7

The Soil Ecosystem

1

Until one hundred years ago, the study of the life in soil was greatly neglected. In the seventeenth and eighteenth centuries, men were far more interested in astronomy than they were in earthworms. Those scientists who were interested in the biological world devoted most of their time to the description and classification of organisms. It was well into the nineteenth century before scientists such as Charles Darwin examined the soil to find out not just what was there, but what was *happening* there. (See *Recommended Reading* 1.) Subsequent investigations by many scientists have revealed a complex community of hundreds of kinds of organisms which directly affects the soil's development.

In the twentieth century the alarming increase in human population has focused attention on the world's food supplies. In spite of the miracles of technology, man still depends, ultimately, on green plants for his food. The green plants, in turn, depend on the soil in which they grow. The importance of that dark quiet world under our feet is indisputable.

Everyone has some knowledge of soil. For some, soil brings to mind the smell of rich black earth in spring. Others see gravel pits eating into hillsides or open cuts beside highways. City people may think of trucks hauling tons of earth out of the excavation for a new skyscraper. Many people think of flower gardens, fishermen think of digging for dew-worms, and children probably think of sandboxes and mudpies.

In all of these associations, soil is only a part of the larger lively environment which is familiar to us, the environment of fresh air and sunshine, water, green plants, animals, and people. But let's change our perspective for a moment. Suppose we are standing on a patch of green lawn in someone's front yard. Now let's shrink, smaller and smaller, until the blades of grass are five times higher than we are. At this level we can see a small hole beside a dandelion plant. Let's crawl in to see what's going on. There is soil very close on every side—in front, behind, above, below. This is quite a different ecosystem from the one we are familiar with. Yet when we look quite closely, we see that it too is very lively. An earthworm inches along just ahead; there are dozens of tiny animals going about their business—eating, finding places to live, caring for their young. With a microscopic eye, we could see still another community consisting of tiny bacteria and algae. The soil is crawling with life. In fact, just a handful of rich, damp topsoil can be home to *billions* of organisms. Where do all those organisms come from? Where do they get their food? Where do they get oxygen to breathe? What happens to them when they die? Maybe we should back out of that hole, go back to normal, and see if we can find some answers.

1.1   WHAT IS SOIL ECOLOGY?

The word *ecology* is Greek in origin. *Oikos* is the Greek word for "household" or "place to live"; *logos* is the Greek word for "thought." Put together, the two words mean, literally, the study of the household. But what can household mean to the inhabitants of the soil? It is necessary to translate the human conception to the environment of soil dwellers. In the soil, household means all of the factors that work together in a particular space. Ecologists refer to such a working unit as an ecosystem. An *ecosystem* is a set of interrelated factors consisting of groups of organisms and their non-living, or physical, environment. *Interaction* is the key word in understanding what is happening in an ecosystem. No factor is independent of the others. Therefore a change in any one factor results in changes in all of the other factors that are directly or indirectly related to it.

The easiest way to understand the nature of interactions between various factors in an ecosystem is to consider the ecosystem most familiar to you. Figure 1-1 illustrates some of the factors that make up the ecosystem of a typical student. The

Fig. 1-1
Some of the interactions which make up the ecosystem of a typical student.

kind of house he lives in (shelter) depends on the climate (weather), the size of his family, and the amount of money the family earns. It may even depend on how wealthy his grandfather was. His part-time job may be mowing lawns, and the hours he works are determined by the weather, by his responsibilities at home and at school, and by social interactions with friends.

These examples of interrelationships are greatly simplified. The human species has developed such a complex social structure that tracing all the interrelated factors in so simple an action as getting up in the morning becomes an impossible task. Try to imagine the complex interactions which occur between cotton plantations and pajamas, between iron ore and alarm clocks, between pine trees and the bedboard, between you as a newborn infant and you now. Enough said, you surely have the idea.

Once you understand how various factors interact under normal conditions, you will understand how one seemingly small change in a complicated picture can have widespread effects. Consider the typical student's ecosystem. A change in any one factor will result in several other changes. For example, suppose the student quits his part-time job. He may now be able to participate in more activities at school; he may have more time to spend with his friends. But his relations with his family may deteriorate because of disputes over money; and he may not have as many new clothes, or be able to spend as much in entertaining his girl friend. Such situations are familiar to everyone.

Although the actual factors are different, the same kinds of interactions occur in the soil. Changes in any one factor may result in changes in many other factors. The size of the springtail population in the soil depends on the temperature of the soil and its moisture content. The number of springtails, in turn, determines how many centipedes live in a certain volume of soil (because centipedes eat springtails). Thus an extended hot dry spell may cause the populations of both springtails and centipedes to decrease.

The more you learn of ecology, the more you will realize that it is very difficult to put your finger on the limit of an ecosystem. A soil ecosystem in nature does not exist independently. Terrestrial animals that feed above ground burrow into the soil to store food or make their homes. Birds such as robins pull earthworms out of the soil. Rain soaks into the soil and sometimes goes through the soil, taking organisms and nutrients with it. We can think of a larger ecosystem which contains all the factors that act outside the soil ecosystem as well as within it. This larger working unit is the world ecosystem. As you can imagine, the interactions in it are extremely complex. However, the same *kinds* of interactions are found even on this very large scale.

Figure 1-2 summarizes the interactions among the five principal factors: climate, plants, soils, animals, and man. Each of these factors affects and is affected by the others. Some of these relationships will be obvious to you. The climate determines the kinds of plants that grow in a particular region. The variety of animal species depends on the food supply. Animal populations can be drastically affected by the presence of human beings. Other relationships may be less apparent. For example, how do animals affect the soil? When you are aware of these interactions it becomes more obvious what effect a change in any one factor would have. If a man decides to cut down all the trees on a piece of land, the soil will be affected. New plants will grow there because of the new physical conditions, and the kinds of animals living on that land will change. Your understanding of the implications of this diagram will increase as you continue your study of soil. The human species has accumulated enough technological skill to be able to cause serious changes in the world ecosystem. One of the most important tasks now facing

Fig. 1-2
The world ecosystem, represented in very simplified terms by the interactions among the five principal factors: climate, soil, plants, animals, and man.

mankind is to learn how to predict the ultimate effect of a change in one factor in the system before an irreversible step is taken.

Let's return to the soil ecosystem. For our purposes we will consider it as a relatively complete set of factors depending on one another. These factors fall into two groups. The first contains all the living organisms, both plant and animal. These are called *biotic* (living) factors. The second group is composed of the non-living factors that contribute to the environment—the kinds of minerals present, the size of the soil particles, the moisture content, the temperature of the soil. All of these are referred to, collectively, as *abiotic* (non-living) factors. There are important interactions between these two groups as well as within each of them, but it is convenient, initially, to consider them separately.

Before we do that, we should take into account the means by which life is maintained. The one absolute necessity for all living organisms is a source of energy. Because energy is so important, we must look carefully at the way it enters the soil ecosystem, and the role it plays there.

1.2   ENERGY IN THE SOIL ECOSYSTEM

The only significant source of energy for living organisms on this planet is the sun. Those organisms which use the sun's energy directly are called *autotrophs* (self feeders); they are able to make their own food. The most familiar autotrophs are green plants. Their green color is due to a pigment, *chlorophyll*, which plays a role in trapping light energy. The energy is combined with carbon dioxide and water to produce energy-rich organic compounds like glucose, a common sugar. The formation of these energy-rich compounds using light energy is called *photosynthesis*. Glucose belongs to a family of compounds known as carbohydrates, which are made of carbon, hydrogen, and oxygen atoms. The hydrogen and oxygen atoms are present in the same proportions as in water, H_2O—hence the name *carbo-hydrate*. Although several carbohydrates are synthesized during photosynthesis, the following word equation best summarizes the main reactions of photosynthesis:

Carbon Dioxide + Water + Light Energy

$$\xrightarrow{\text{chlorophyll}} \text{Glucose} + \text{Oxygen}$$

This process occurs principally in the leaves of green plants. The glucose is then transported to various locations in the plant where energy is needed. It also serves as a starting material for the synthesis of proteins, fats, and other organic compounds that the plant requires.

Plants need energy for many purposes—growth, replacement of worn-out tissue, absorption of water and nutrients, reproduction, and even for photosynthesis! So some of the carbohydrates are broken down again at sites where energy is needed. When the carbohydrates are broken down, the energy stored during their formation is released.

Every time the form in which energy is stored changes, some energy is lost. The least waste occurs when carbohydrate molecules are broken down in the presence of oxygen. This means of energy release is called *aerobic respiration*. The process may be summarized by this word equation:

$$\text{Glucose } + \text{ Oxygen} \rightarrow \text{Carbon Dioxide } + \text{ Water } + \text{ Energy}$$

Compare this equation with the one for photosynthesis. Notice that the oxygen used in respiration was produced by green plants as a by-product of photosynthesis.

A second means of energy release is *anaerobic respiration*. This process occurs when oxygen is not available. More energy is wasted, but it is a valuable alternative in the soil, where oxygen supplies can be used up.

But how is this related to the soil ecosystem? Living organisms which cannot trap the sun's energy directly are called *heterotrophs* (other feeders). They feed on other organisms to get the energy-rich food they need. The food is digested and stored as energy-rich organic compounds useful to them. Each time stored energy moves from one organism to another, some of it is wasted. This does not mean that it disappears, but rather that it changes into a form which is not useful in carrying on life processes. Usually this wasted energy takes the form of heat which radiates from the organisms.

All heterotrophs depend, ultimately, on autotrophs to convert light energy from the sun to chemical energy held in organic compounds. Almost all soil organisms are heterotrophs. Let us see how they get the energy they must have to continue to live.

Fortunately for the heterotrophs, the autotrophs convert, or *fix*, much more light energy than they need. And, as a by-product, they release much more oxygen than they require for respiration. That energy and that oxygen maintain the soil community. Parasitic nematodes (hair-like roundworms) and root-feeding insects obtain energy by eating the roots of green plants. Fallen leaves are pulled underground by such organisms as earthworms and millipedes. These leaves, along with other dead plant and animal material, constitute *litter*. It provides one of the largest sources of energy in the soil. Other organisms do not eat green plants directly but eat the heterotrophs that eat the green plants. The excess oxygen resulting from photosynthesis filters through the air spaces in the soil. It is used by heterotrophs for aerobic respiration—to break down the energy-rich molecules and release the energy.

Figure 1-3 summarizes how energy flows from organism to organism in the soil ecosystem. As the figure shows, energy is distributed for various purposes at each *trophic* (feeding) level. In a plant, some of the stored energy is used for the plant's own life processes. Further energy is stored in leaves, stems, and roots. Thus a root-feeding nematode gets only a small quantity of the total energy that the plant fixes. Of this energy acquired by the nematode, some is used in maintenance, movement, and other body functions. Some is in forms not useful to the nematode and is excreted. The material excreted is used as a source of energy by such microorganisms as bacteria and fungi.

Fig. 1-3
When stored energy moves from one organism to a second, only a fraction of the energy transferred is stored in the body tissue of the second organism.

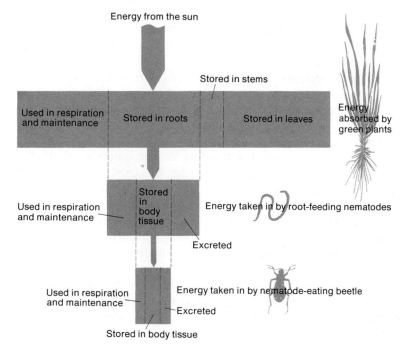

Some of the energy taken in by the nematode is stored in energy-rich organic molecules as part of its body weight. Only this portion can be transferred to the next heterotroph, a nematode-eating beetle. Note that only a small part of the energy initially fixed by a green plant reaches the beetle. Note, too, that the longer the chain, the less energy the organism at the end gets.

The need for energy is one of the chief reasons for the interactions between organisms. In the next section, these relationships are examined in more detail.

1.3   BIOTIC ASPECTS OF THE SOIL ECOSYSTEM

The best way to begin a study of the living organisms in a soil ecosystem is to do some digging in moist, fertile soil with a shovel or a trowel. The digger must be alert, for the disturbed soil animals very quickly disappear. The very fact that the environment is disturbed means that the situation observed will not be the normal one. The animals will try to escape the abnormal situation of being exposed to the sun and open air, rather than being underground. Another problem in the study of soil organisms is the very small size of many of them. The human eye just isn't powerful enough to observe bacteria and protozoa. Many of the fungi are very small and almost transparent. Even larger organisms such as mites and nematodes are almost impossible to see against the dark brown, irregular background of the soil.

The student of ecology wants to learn how these organisms affect each other under normal conditions. Such a study has occupied hundreds of ecologists for dozens of years. They have attempted to observe soil animals in "normal" activity. One of the most fascinating accounts of such research is described in *Recommended Reading* 2. The outcome of all this work has been the realization that, for the most part, the same basic interactions occur in all soil ecosystems. In fact, all ecosystems (soil, grassland, forest, pond, and so on) have essentially the same interactions. The only differences are in the kinds of organisms that play the various roles. The following is a brief summary of these basic interactions as they occur in the soil.

As you saw in Section 1.2, all living organisms must have a continuous supply of energy to stay alive. Further, most of the living organisms in the soil environment are heterotrophs, depending on other organisms for their energy. Thus, it cannot

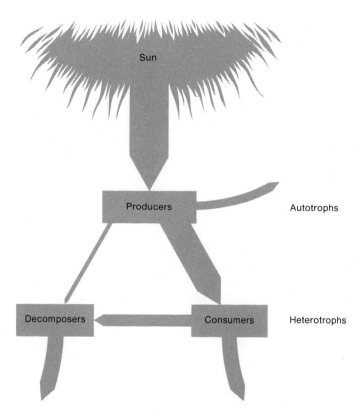

Fig. 1-4
Schematic representation
of energy flow in the soil
ecosystem.

be said that the soil ecosystem is completely self-sufficient.
Energy is fixed by autotrophs outside the ecosystem and is
brought into it by various routes. Green plants store carbohydrates in roots; leaves that fall off green plants are pulled
down into the soil by litter-feeders. Figure 1-4 shows how energy
reaches the various living organisms in the soil.

The terms *producer*, *consumer*, and *decomposer* are
generally used to describe the basic roles played by organisms in
an ecosystem. Autotrophs produce energy-rich organic molecules
using light energy, and therefore are called *producers*. There are
two kinds of heterotrophs—consumers and decomposers. *Consumers* eat organic material and digest it internally. *Decomposers* (microorganisms such as bacteria and fungi) get their
energy by releasing enzymes into dead plant and animal
material. When the organic material is partially broken down,
the decomposers absorb the smaller organic compounds.

From each box in Figure 1-4 an arrow points to
"nowhere." These arrows represent the energy lost in the

transfer from organism to organism and the energy used by the organism for movement and other life processes. It is important to note that energy moves in only one direction and is gradually used up. The sun must always be supplying more.

This diagram represents, in simplified terms, the interactions which occur in any ecosystem. Every living organism plays at least one of the three roles. (A few organisms are able to handle two roles. Can you name some?) A horizontal line, to represent the surface of the soil, could be drawn immediately below the box labeled "Producers."

Now, let us study each role and each interaction in more detail. A sometimes frustrating fact of biology is that for every generalization, there is usually an exception. For example, producers grow above ground—with one exception. One group of algae occurs in loose soil to depths as great as 20 centimeters and can make their own food by means of photosynthesis. Generally speaking, however, the producers which fix most of the energy that sustains the soil ecosystem are the green plants which photosynthesize above ground.

The role of consumer is played by a large number of species which, at first, seem to have very little in common. Some species eat the producers. These are called *herbivores* (meaning plant-eaters) or *primary consumers*. Root-feeding insects and parasitic nematodes are examples of such animals. The amount of energy obtained by soil organisms through the consumption of plant detritus far exceeds that obtained by the consumption of living plants. The abundance of earthworms and other litter-feeders in fertile soil is evidence of this fact. (*Plant detritus* refers to dead organic fragments such as dead leaves, twigs, and algae.)

The next group of consumers are the *carnivores* (meaning flesh-eaters). *Primary carnivores* (also called *secondary consumers*) get their energy by eating herbivores. A spider that eats a springtail is a primary carnivore. In some cases *secondary carnivores* (or *tertiary consumers*) may be present and prey on the primary carnivores. For example, the robin that ate the spider that ate the springtail that ate the algae, is a tertiary consumer. In other words, it is an animal that eats animal-eaters.

Animal detritus is also an important factor in energy flow in the soil. It consists of fragments of dead animals and excreta from animals. (Excreta is the organic material which has been eaten but cannot be used by the eater, and hence is excreted.) Most of the organisms that eat plant detritus are equally effective at eating animal detritus. In fact, they generally do not distinguish between them; detritus is detritus. It is only because of man's tendency to organize and pigeon-hole facts that any problem arises. If you can handle the idea that an earthworm can be both a herbivore and a carnivore in the same bite, then there really is no problem.

Some specialized consumers are almost always present in soil ecosystems and should be mentioned here. A *parasite* feeds on another living organism but does not injure it sufficiently to kill it. Both animal and plant parasites exist, and almost every living organism is susceptible to at least one of them. In the soil, most parasites are bacteria and protozoa, but the group also includes nematodes and insects. *Saprophytes* are non-green plants that get their energy by digesting the organic material of dead plants and animals. Fungi such as molds and yeasts are the main saprophytes of the soil ecosystem. *Scavengers* are animals that get their energy by eating dead plants and animals. Litter-feeders like earthworms are scavengers.

The third principal group of organisms are the *decomposers*. They are extraordinary microorganisms that do the job of digestion outside themselves. They release digestive enzymes which break down organic material into smaller molecules. They then absorb the digested material. These small organic molecules supply sufficient energy to meet the decomposers' needs. Bacteria and fungi are the two main kinds of decomposers. These are the organisms that occur in such astronomical quantities in soil ecosystems. Of the billions of organisms in a handful of soil, 99.99% are minute decomposers.

A special group of decomposers deserves particular attention. It consists of the *transformers*. They break down simple organic molecules into molecules that are small enough to be absorbed through the roots of producers. This final step in the breakdown of organic molecules is critical to the survival of an ecosystem. Under normal conditions, there is no continual input of essential elements. The amounts of such elements as nitrogen and phosphorus are, relatively speaking, fixed. It is therefore necessary to recover those elements from the dead organic tissue so they can be used again. The transformers provide the closing link in the cycle by releasing the elements in a form useful to green plants. The cycling of essential elements (or nutrients) is illustrated by the black arrows in Figure 1-5 (page 14). *Nutrient cycles* are so important to the maintenance of any ecosystem that Section 1.4 is devoted to a careful look at three of them.

The most important activity of soil organisms is acquiring energy. As a result, the chief interactions between species are "eating" and "being eaten." The energy flow can be followed through a succession of organisms that "eat and are eaten." A list of such interrelated organisms is called a *food chain*.

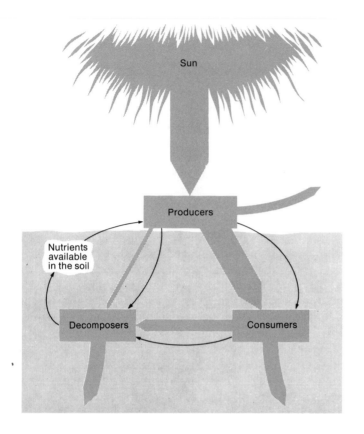

Fig. 1-5
Energy flow and nutrient cycling in the soil eco-system.

Fig. 1-6
A food chain: fallen leaf → springtail → centipede.

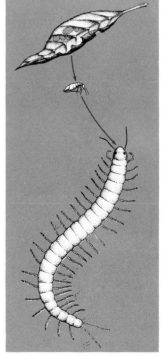

Leaf → Springtail → Centipede

is one food chain that is common in soil ecosystems (Fig. 1-6). Generally food chains follow the pattern,

Producer → Herbivore → Carnivore

When several food chains are drawn to represent various interactions in an ecosystem, we find that many species participate in more than one food chain. Such multiple roles are common. Ecologists have coined the term *food web* for a set of interconnected food chains (Fig. 1-7).

The particular role played by a species in an ecosystem is called an ecological *niche*. The niche occupied by a pseudo-scorpion in the soil ecosystem is that of a carnivore. The earthworm occupies the niches of herbivore, carnivore, and scavenger. An interesting explanation of the term niche is found in *Recommended Reading* 3.

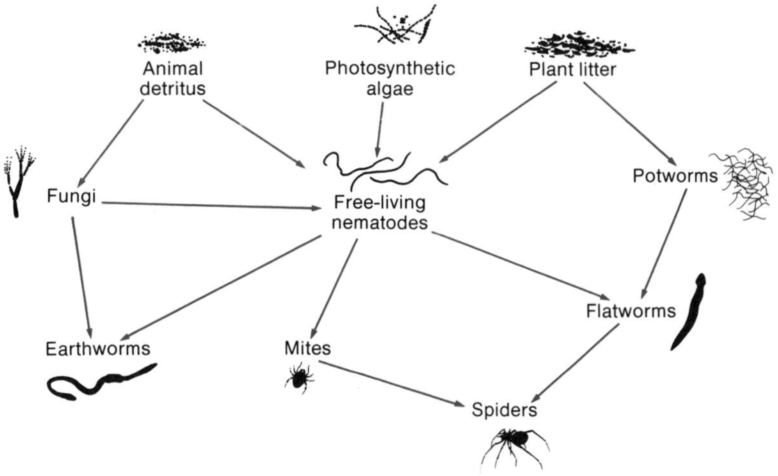

Fig. 1-7
Interrelated food chains make a food web. How many food chains can you find in this food web?

1.4 NUTRIENT CYCLES IN THE SOIL ECOSYSTEM

Carbon, hydrogen, nitrogen, and oxygen could be called the building blocks of life. Every living organism is composed almost entirely of these four elements. In addition, lesser quantities of more than a dozen other elements must be present for certain life processes to occur. For example, magnesium is an essential element for green plants because it is a constituent of chlorophyll. Photosynthesis cannot occur without it. Red-blooded animals need iron for the hemoglobin molecule which carries oxygen in the bloodstream.

Because of their importance to living organisms, elements like carbon, hydrogen, nitrogen, oxygen, magnesium, and iron are commonly called *nutrients*. A forest, a grassland, or a pond can maintain itself without the addition of nutrients by man, provided man removes nothing from the ecosystem. Obviously, then, nutrients are recycled in ecosystems. To illustrate how this happens, three nutrient cycles are described. Particular attention is given in each case to the movement of the nutrients *within the soil*.

The Water Cycle (Hydrologic Cycle). Hydrogen and oxygen are available in every ecosystem in the form of water. The movement of water in a soil ecosystem is summarized in Figure 1-8. Water can enter the soil by two means—soaking down from the surface, and moving up from deeper regions where ground water is present. The movement down is called *percolation* and the movement up, *capillarity*. The surfaces of soil particles are usually covered with a fine film of water. This

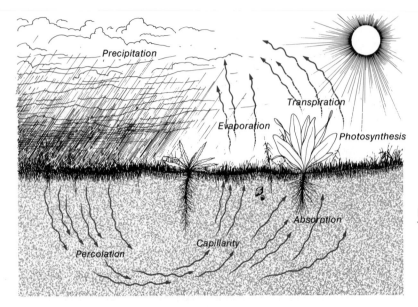

Fig. 1-8
The water cycle.

provides a suitable environment for such organisms as free-living nematodes, flatworms, protozoa, and bacteria. The humidity in the air spaces between the soil particles is usually close to 100%. Therefore the water films do not evaporate and, as a result, the soil organisms do not lose much water by evaporation. Green plants are continually taking in water. Part of it is transpired (released through pores in the leaves) and part of it is used in photosynthesis. As you can see from the diagram, it is necessary to go outside the soil ecosystem to complete the cycle. Remember that, under natural circumstances, a completely self-sufficient soil ecosystem does not exist.

The Carbon Cycle. Like the water cycle, the carbon cycle is not complete within the soil. The largest quantity of carbon available to the cycle is in the oceans as dissolved carbon dioxide. As a result, the oceans are mainly responsible for maintaining a carbon dioxide balance in the atmosphere. Excess carbon dioxide in the air dissolves in the oceans, and deficits of carbon dioxide in the air are replaced by carbon dioxide from the oceans. Through the combustion of fossil fuels, man is gradually throwing this part of the carbon cycle out of balance. Carbon dioxide is being added to the air faster than the oceans can absorb it.

Figure 1-9 illustrates the principal pathways of carbon through the soil, and into and out of the atmosphere above the soil. Green plants use carbon dioxide in the photosynthesis of carbohydrates. Both plants and animals release carbon dioxide

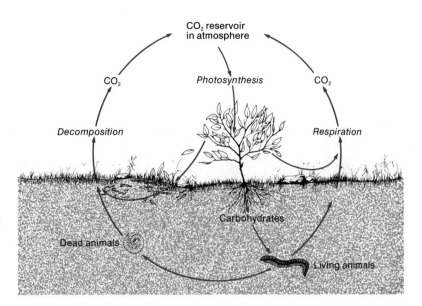

Fig. 1-9
The carbon cycle in and above the soil.

in respiration. When carbohydrates are consumed by soil herbivores, the carbon is either incorporated in the animal tissue, respired, or excreted. Decomposers work on excreted organic matter and dead organisms, releasing carbon dioxide.

You may have noticed from the diagram that carbon is always associated with oxygen. This cycle, therefore. is an oxygen cycle as well. To be complete as an oxygen cycle, the diagram would have to be somewhat modified. Arrows would be added to represent production of oxygen as a by-product of photosynthesis, and consumption of oxygen by plants and animals in respiration. The process of decomposition also uses up oxygen.

The Nitrogen Cycle. The soil plays an important role in the nitrogen cycle. Certain bacteria and fungi act on various nitrogen-containing substances in assembly-line fashion, gradually changing them to forms suitable for re-use by green plants.

Follow Figure 1-10 as you read this description of the nitrogen cycle. About 80% of the air we breathe is nitrogen. However, the form in which nitrogen exists in the air is of no use to most organisms. Fortunately, certain kinds of bacteria can cause atmospheric nitrogen to react with oxygen to produce nitrates. They are called nitrogen-fixing bacteria. Nitrates can be absorbed by plants. They are then used in the synthesis of amino acids (the building blocks of proteins) and other nitrogen-containing organic substances. Animals obtain the nitrogen they require by eating plants or by eating animals that eat plants. It

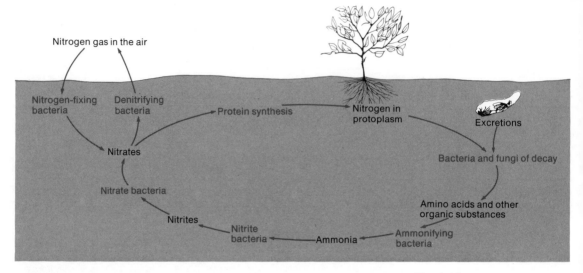

Fig. 1-10
The nitrogen cycle in the soil.

might seem that nitrogen movement would be halted at this point, with the nitrogen tied up in complex organic substances. Although large amounts of nitrogen are stored at this point in the cycle, a net movement back to simpler forms does occur. Animal excreta and dead organisms are broken down into amino acids and simple organic substances by bacteria and fungi of decay. These smaller molecules are then transformed by the action of another kind of bacteria, ammonifying bacteria, to ammonia. The ammonia is changed to nitrite by nitrite bacteria, and the nitrite to nitrate by nitrate bacteria. Nitrates are then absorbed by plants to complete the cycle.

Another group of soil organisms should be mentioned—denitrifying bacteria. These organisms take in nitrogen in the form of nitrates and release nitrogen molecules (such as occur normally in the air). By so doing, they complete the branch of the nitrogen cycle involving atmospheric nitrogen.

The interaction between biotic and abiotic factors illustrated in the nitrogen cycle is a good example of the interdependency within an ecosystem. For example, try to visualize the effects of destroying the nitrite bacteria in the soil of a field.

1.5 ABIOTIC ASPECTS OF THE SOIL ECOSYSTEM

The non-living factors within a soil ecosystem are called *abiotic* factors. A host of chemical and physical factors fall into this category.

Foremost among the chemical factors are the elements that compose the inorganic part of the soil. Most of this inorganic material originates from the parent material which makes up the earth's crust. Through the action of climate and living organisms over tens of thousands of years, rock is gradually broken down into smaller and smaller fragments. The composition of these fragments affects a soil ecosystem. About 20 different elements are essential for life. For example, some of the organisms which fix atmospheric nitrogen must have trace quantities of molybdenum. Yet essential elements are harmful if they are present in too high concentrations. Lawn fertilizer is a source of three essential elements—nitrogen, phosphorus, and potassium. Without these elements grass cannot grow. Yet you have probably seen a lawn that has been "burned" by the application of too much fertilizer.

Soil contains simple organic molecules resulting from the action of decomposers on animal excreta and dead tissue. These are considered abiotic parts of the ecosystem. As you would expect, they interact with other factors. As an illustration, consider amino acids in the soil. The abundance of amino acid molecules depends on the activity of bacteria and fungi of decay, which, in turn, depends on the amount of dead plant and animal tissue in the soil. Since organisms are constantly dying, the concentration of amino acids would increase indefinitely were it not for the ammonifying bacteria which break them down and release ammonia.

Water and the atmospheric gases—nitrogen, oxygen, and carbon dioxide—all fall into the abiotic category. The importance of these factors need not be stressed again. Section 1.4 describes their movements through the soil ecosystem and their principal interactions with other factors.

The abiotic factors which have been mentioned are chemical in nature—elements in rock particles, simple organic molecules, water, and atmospheric gases. Other abiotic factors are physical in nature.

The size of rock fragments in soil is important. Soil of very small particle size has fewer and smaller spaces to provide shelter for soil organisms. The particle size also affects the movement of water through the soil and the growth of fungi and the roots of green plants.

Energy from the sun enters the soil in two forms, as heat and as light. Heat energy affects the soil temperature which, in turn, influences the rate of metabolism (life processes) of the living organisms. The nature of the soil environment results in very little variation in temperature as compared to variations above ground. Just which soil organisms inhabit a particular location depends to some extent on what temperature range exists there.

Light acts indirectly on the soil through photosynthesis by green plants above the ground. More directly, light which penetrates a few centimeters into the soil permits photosynthesis by certain algae (see Section 1.3).

Even the wind can exert an influence on the soil. If a strong wind blows on a dry area with little vegetation, it may move away surface layers of the soil. Formerly covered parts are then exposed. Sand dunes along large lakes are particularly susceptible to such "blowouts" since the vegetation is usually sparse, the soil dry, and the wind strong in such an area (Fig. 1-11). Overnight a wind storm can change a forested dune into a stark desert. And hundreds of years may pass before nature can rebuild the forest. Today foresters plant trees and shrubs on sand dunes to prevent blowouts. Should a blowout still occur, the planting of selected species of trees and shrubs speeds up the return of the area to a forest (see *Recommended Reading* 5).

The abiotic factors introduced here are the most important ones in a soil ecosystem. Unit 2 describes some of them in

Fig. 1-11
A blowout in a sand dune region on the northern shore of Lake Ontario. (Courtesy of Ontario Ministry of Natural Resources.)

greater depth. Sometimes the effects of abiotic factors are difficult to measure accurately, but they should always be considered during your field studies.

For Thought and Research

1 Scientists of the eighteenth century were more interested in the moon and the stars than they were in soil.

(a) Why do you think this was the case?

(b) Has this tendency changed in men of the twentieth century? Why?

(c) Which do *you* think is more important? State your reasons. (This topic could be used for a class debate.)

2 (a) Explain the meaning of the terms "ecology" and "ecosystem," referring to the soil in your school yard to illustrate your answer.

(b) What do you consider to be the most important aspect of all ecosystems? Why?

3 Why is it difficult to isolate a soil ecosystem and have it continue to exist relatively unchanged?

4 Make two lists of the factors which have affected you today. In the first, put biotic factors; in the second, put abiotic factors.

5 The poet Pope took a rather strong stand concerning the balance in nature when he expressed the following in his "Essay on Man" in 1733:

> From Nature's chain whatever link you strike
> Tenth, or ten thousandth, breaks the chain alike.

Do you agree or disagree with his statement? Whatever your stand, support your view with logical reasoning, using evidence such as personal experiences.

6 Using Figure 1-1 as a model, create a diagram to represent the interactions that you think might occur in an earthworm's ecosystem.

7 Figure 1-12 represents the interdependence which exists between a few of the factors in a soil ecosystem. Write an explanatory sentence or two for each relationship that has been designated by a letter (a, b, c . . . i).

Fig. 1-12
Some of the interactions in a soil ecosystem.

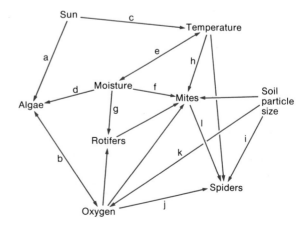

8 (a) Why do all living organisms need energy?

(b) What is the principal source of energy for all living organisms on this planet?

(c) What is the major means of getting energy into the soil?

(d) What prevents energy from being recycled like nutrient elements?

9 Most soil organisms are heterotrophs.

(a) What does "heterotroph" mean?

(b) Why are most soil organisms heterotrophs?

(c) What term is used to describe the group of organisms which are self feeders?

(d) What soil organisms belong in the group described in (c)? What niche do they occupy in the soil ecosystem?

10 (a) Could an ecosystem survive without secondary and tertiary consumers? Explain your answer.

(b) Why are decomposers essential for the maintenance of a soil ecosystem?

11 (a) What do scavengers and saprophytes have in common?

(b) How do they differ?

(c) Name a third group of soil organisms which perform a similar function.

12 (a) What niches are occupied by yeasts and molds in the soil ecosystem?

(b) Name five other organisms that occupy more than one niche. For each, name the niches occupied. Your examples need not be from the soil ecosystem.

13 The following food chains exist in most soil:

(a) plant detritus → earthworm

(b) plant detritus → bacteria → springtail → centipede

(c) animal detritus → bacteria → protozoa → flatworm → beetle

(d) wood → termite → centipede

Using Figure 1-7 as a model, try to connect these chains to make a food web. Account for any problems you encounter.

14 Suppose you could "tag" a particular carbon atom in a carbohydrate molecule in a plant. Write a short description of what *might* happen to that carbon atom when the plant is eaten by a soil herbivore. Continue the account until the atom becomes part of a new plant.

15 Figure 1-13 shows how the amount of water in the soil affects the rate at which ammonia is converted to nitrate.

(a) What is the optimum moisture content for conversion of ammonia to nitrate?

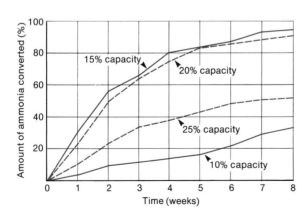

Fig. 1-13
Conversion of ammonia to nitrate in soils containing different amounts of water.

(b) Suggest reasons why the process occurs more slowly when the soil contains 25% of the moisture it can hold than when the soil contains 20%.

(c) Predict where the line would be drawn to represent conversion of ammonia to nitrate when the soil contains 100% of the moisture it can hold. Explain the reasoning behind your prediction.

16 Figure 1-14 shows how the proportion of oxygen in soil changes over the period of a year. It also shows how the oxygen content is affected by the composition of soils.

Fig. 1-14
Annual variation in oxygen content of soil gas for two soils.

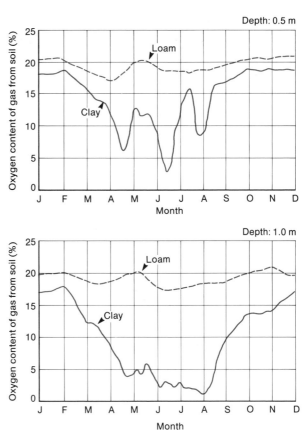

(a) What processes use up oxygen?

(b) Why does the amount of oxygen decrease during the summer months?

(c) How is the oxygen replaced?

(d) Predict the shape of the lines drawn to represent the amount of oxygen at a depth of 2 meters. Defend your prediction.

(e) Suggest a reason why the variation is less in loam than in clay.

(f) Do you think there would be more organisms in the loam or in the clay? Give reasons for your answer.

Recommended Readings

1 *The Formation of Vegetable Mould Through the Action of Worms* by Charles Darwin, 1881. This book was reissued in 1945 by Faber & Faber under the title *Darwin on Humus and the Earthworm.* Consult it for information regarding the earliest studies of the soil ecosystem.

2 *Soil Animals* by F. Schaller, University of Michigan Press, 1968. Consult this book for an interesting account of research into the normal activities of soil animals.

3 *Basic Ecology* by Ralph and Mildred Buchsbaum, Boxwood Press, 1957. Read pp. 61-62 for a clear explanation of the term "ecological niche."

4 *Concepts of Ecology* by E. J. Kormondy, Prentice-Hall, 1969. Consult this book for more detailed descriptions of nutrient cycles.

5 *A Guide to the Study of Terrestrial Ecology* by W. A. Andrews et al., Prentice-Hall, 1973. See Section 1.5 for a description of plant succession on sand dunes.

6 *Ecology of Soil Animals* by J. A. Wallwork, McGraw-Hill, 1970. This book contains thorough, accurate, and detailed descriptions of all aspects of soil ecosystems.

Origin and Nature of Soil

2

Soil is more than a lifeless layer of "dirt." It is also more than a thin layer of minerals built up over a long period of time to hold a supply of things necessary for plant growth. Soil is a dynamic layer in which many complex chemical, physical, and biological activities are going on constantly. The soil determines the organisms in it, the organisms modify the soil, and the modified soil can then support a different population of organisms. The soil is a changing body, adjusting to conditions of climate, topography, and vegetation. Because of this dynamic interaction between living and non-living components, soil studies are exceedingly interesting. For the same reason, soil studies can become quite complex. In this Unit you will find out what soil is, how it is formed, and why it differs from place to place.

2.1 WHAT IS SOIL?

To different people the word "soil" has different meanings. To you, perhaps, soil is the lumps in your father's garden, the material under the grass of your school's lawn, or the furrows in a farmer's field. To the farmer, soil is a medium for plant growth. To the student of *pedology*, the study of soils, it is the fertile surface material of the earth which is capable of supporting plant growth. How thick do you suppose the soil is in relation to the overall size of our planet? Imagine a giant tomato with a diameter of 200 feet and a skin thickness the same as that of an ordinary tomato. The thickness of the skin on the giant tomato bears the same relationship to the size of the tomato as the thickness of the soil does to our earth.

For proper plant growth, soil must be made up of substances in three states—solid, liquid, and gas. The solid portion of soil is both inorganic (rock fragments) and organic (plant and animal materials, both living and dead). The liquid portion is a complex solution of chemical compounds necessary for many important activities in the soil. The gases are basically those found in the atmosphere, together with gases liberated by biological and chemical activity in the soil. They are found in the open pore spaces of the soil.

Inorganic Components. Rocks are the source of the inorganic substances in soils. These substances are normally rock fragments and minerals of various kinds. The fragments are remnants of larger rocks and are usually quite coarse. Table 1 compares the major size classes of inorganic particles.

TABLE 1

Size	Diameter (mm)	Visible using
Gravel	Greater than 2.0	Naked eye
Coarse sand	2.0–0.2	Naked eye
Fine sand	0.2–0.02	Naked eye
Silt	0.02–0.002	Light microscope
Clay	Less than 0.002	Electron microscope

The minerals may be as large as the smallest rock fragments or too small to be seen with the aid of an ordinary microscope. *Primary minerals*, such as quartz, have remained unchanged in composition from the original rock. Other minerals, called *secondary minerals*, have been formed by the weathering of less-resistant minerals over a long period of time. Generally speaking, primary minerals tend to be coarse-grained, while secondary minerals, especially those in clays, are predominately fine-grained. Why do you suppose that this is the case?

Organic Components. The non-living organic matter in soil is an accumulation of dead, but intact, plant and animal tissue together with partially decomposed tissue. In most mineral soils the amount of organic matter usually ranges from less than 1% to about 5% by weight. Desert soils, for example, commonly contain much less than 1% organic matter by weight. In contrast, some peat soils have an organic content close to 100% of the dry weight.

The organic matter is mainly responsible for black and brown colors in soil (Fig. 2-1). It is also largely responsible for the loose condition of productive soils, and is thus very important to the soil. It is the major soil source of phosphorus and sulfur, two important nutrient elements. In addition, it is the sole source of nitrogen. Organic matter provides the main energy source for many soil microorganisms. It also helps to increase the water-holding capacity of the soil.

Green plants growing on the soil provide most of the organic matter present in soil. As these plants die, a layer of debris consisting of leaf litter, plant tops, and roots accumulates. Centipedes, millipedes, earthworms, springtails, mites, and many other soil animals eat this debris and break it down into simpler organic compounds. These compounds are present in the excreta of the animals. Microorganisms like bacteria and fungi further decompose the organic compounds as they feed on the excreta. Finally, most of the organic matter is changed into carbon dioxide and water, the two compounds which were originally

Fig. 2-1
Note the differences between (A) a soil low in organic content, and (B) a soil high in organic content.

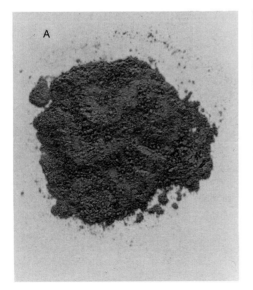

used by the green plants to make the organic matter. However, some of the organic matter does not decompose easily. As a result, it remains in the soil for long periods of time. This more resistant organic matter is known as *humus*. It is generally black or brown in color.

Both soil temperature and soil aeration influence the rate of decomposition. Organic matter tends to accumulate in soils which lack aeration, since aerobic respiration is reduced. Similarly, the organic content of soil increases in cooler areas due to the reduced activity of microorganisms. Thus warmer climates tend to produce soils that are lower in organic matter than those in colder climates.

Soil Solution. The spaces between the particles of soil, the *pore spaces*, can be occupied by either water or air. In any given soil the proportions of water and air are interrelated; as one increases, the other decreases. Let us deal first with the *soil solution*. This is the water held within the soil pore spaces, together with its dissolved materials. All plants need water for growth. It is important, therefore, to know how water moves in the soil, how much moisture the soil can store, and how much of this moisture is available to the plants. These factors are determined, as you might expect, by the size and distribution of the pore spaces in the soil and by the attraction of the soil particles for moisture.

Some water is retained by the soil after a rain, even though there has been every opportunity for the water to drain off. This water completely fills small pores in the soil as it moves downward under the influence of gravity. However, many of the pores in the soil are so small that they serve as capillaries, allowing the water to move against the pull of gravity. Water moving in this fashion is known as the *capillary water*. You can demonstrate *capillarity* by placing pieces of glass capillary tubing, each with a different diameter, in a beaker of water. Hold the tubes upright and at various angles (Fig. 2-2).

Another form of soil water occurs as an extremely thin film on the soil particles. It is called *hygroscopic water*. This water cannot move as a liquid since it is tightly attracted to the soil particles. Also, a small portion of the soil water, called *combined water*, is chemically bound with soil materials. Neither of these forms makes the soil "wet," and intense heat is needed to drive them from the soil. You can observe combined water by heating some crystals of copper(II) sulfate or magnesium sulfate in a test tube.

If you examined some soil after a period of heavy rainfall, you would notice that all of the soil pores were

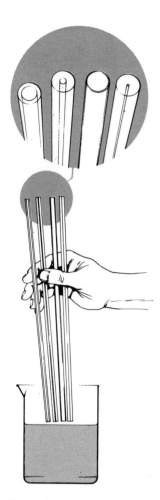

Fig. 2-2
Lowering these tubes into the water will illustrate capillary action. Can you account for the results?

completely filled with water. This occurs because the water has moved downward in the soil, replacing the air in the soil spaces. With continued rainfall, this *gravitational water* will move farther and farther into the soil, carrying dissolved minerals with it. If enough water falls, it will eventually move into the underground water table.

After all the gravitational water has drained away, the soil is said to be at *field capacity*. This field capacity is made up of the capillary, hygroscopic, and combined water, plus any water vapor that is present. As the gravitational water drains away, the water in the larger pore spaces is replaced to a certain extent by air. However, the smaller pore spaces are still completely filled with water.

As living plants draw water from the soil, the amount held in the capillary spaces is reduced. The plants find it more and more difficult to get water. Finally the supply becomes so low that the plants wilt. Unless water is soon added, the plants will not recover from wilting. The moisture content of the soil at this point is called the *permanent wilting percentage*. The amount of water which can be obtained by the plant, the *available water*, is thus the difference between the field capacity and the permanent wilting percentage. The water that cannot be obtained by the plant is called the *non-available water*. It is made up of the hygroscopic water, the combined water, a fraction of the capillary water, and the water vapor. Figure 2-3 summarizes the forms of water and their availability to plants.

Soil Air. Soil contains air in addition to organic matter, inorganic matter, and water. This air is found in the pore spaces

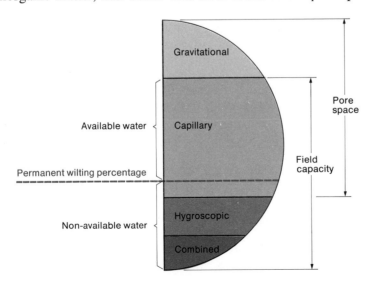

Fig. 2-3
Forms of soil water.

of the soil. Soil air differs from atmospheric air in several ways: it is not continuous since it is separated by soil particles; it generally has a higher moisture content; its carbon dioxide content is usually higher and the oxygen content lower than in atmospheric air. (The carbon dioxide and oxygen content of the soil depend mainly on the biological activity in the soil.) Table 2 compares the percentages of the main gases in the atmosphere and soil.

TABLE 2 SOIL AIR COMPOSITION

| Gas | Percentage | |
	Atmosphere	Soil
Oxygen	20.99	20.3
Carbon dioxide	0.04	0.5
Nitrogen and other gases	78.97	79.2

The amount of air retained by soil depends to a large extent on the amount of water in the soil. The air moves into those soil pores not occupied by water. Since the soil water vacates the larger pores first, these tend to fill first with air. As the soil dries out, the air occupies pores of smaller size. Thus soils with a high percentage of small pores tend to be poorly aerated. Clay soils are in this category. Since oxygen in the soil is necessary for good plant growth, such soils are unsatisfactory for most agricultural purposes. For the same reason, soils dominated by water are also unsatisfactory.

As you have seen, soil is the weathered surface layer of the earth's crust, intermingled with living organisms and the products of their decay. Thus a representative soil consists of

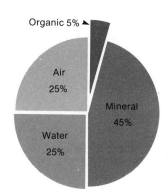

Fig. 2-4
The approximate percentages by volume of water, air, organic matter, and inorganic (mineral) matter in an average soil.

1) a mineral fraction (modified parent material or inorganic foundation);

2) an organic fraction (both dead organic material and living organisms);

3) pore spaces between the soil particles containing air and water.

The approximate percentages of each component are given in Figure 2-4. These percentages do, of course, vary from place to place.

For Thought and Research

1 Soil is an ever-changing body of matter which responds to several influences. Name as many of these influences as you can and explain clearly how each has an effect on the soil.

2 (a) Why does organic matter impart a brown or black color to soil?

(b) Why does organic matter increase the water-holding capacity of soil?

(c) Experts on lawn care recommend that grass clippings be left on the lawn. Why?

(d) What relationships might exist between soil color and climate? Why?

3 (a) Find an explanation for capillarity.

(b) Why do gardeners and farmers keep the top few centimeters of soil in a loose condition?

4 (a) Why is the field capacity the sum of the capillary water, hygroscopic water, combined water, and water vapor?

(b) Explain why the available water is the difference between the field capacity and the permanent wilting percentage.

(c) Do you think the permanent wilting percentage will be different for various species of plants? Why?

5 (a) Explain how biological activity affects the oxygen and carbon dioxide content of soil.

(b) Why is aeration of soil important to plant growth?

(c) Why might excessive watering harm plants?

(d) Why will a soil with abundant organic matter likely provide better aeration than a clay soil?

6 Perform the field and laboratory studies in Sections 5.3-5.11 and Sections 5.16-5.19.

Recommended Readings

1 *Elements of Ecology* by G. L. Clarke, John Wiley & Sons, 1966. This book contains a good general discussion of soils.

2 *Introductory Soils* by K. C. Berger, Macmillan, 1969. The first 100 pages are a good source of additional information on soil water and organic matter.

3 *Soil Conditions and Plant Growth* by E. W. Russell, John Wiley & Sons, 1963. A comprehensive book on all aspects of soil science.

4 *The Nature and Properties of Soils* by H. O. Buckman and W. C. Brady, Collier-Macmillan, 1970. An interesting but detailed account of soil science.

2.2   SOIL FORMATION

Now that you have a general idea of what soil is, we can go one step farther to find out how it is formed. Perhaps you do not know that the soil in your area is not the same as that in some other parts of North America. Soils differ from country to country and, in some cases, even from town to town. The soil in

Alaska is very different from that in Ontario, and the soil in Colorado is unlike that in Saskatchewan. What causes these differences? What factor or factors work to make soils different from place to place? Before we can discuss these factors in detail we must look at the *parent material* from which the soil is formed.

PARENT MATERIAL

The parent material consists of rocks which have been broken down. The earth's rocks fall into three main categories— igneous, sedimentary, and metamorphic—according to their method of formation. *Igneous rocks* are often referred to as primary. The others are called secondary because they are formed largely as a result of changes in the primary rocks. Igneous rocks (Fig. 2-5) are formed from molten material which exists under conditions of high temperature and pressure deep

Fig. 2-5
Some igneous rocks. What evidence do you see that granite and gabbro formed by cooling underground, while pumice and obsidian formed by cooling above ground?

Granite

Gabbro

Pumice

Obsidian

Sandstone

Shale

Halite

Fig. 2-6
Some sedimentary rocks.
Find out how each of these
rocks formed.

within the earth. Rocks are produced as this molten material cools. If the cooling takes place very quickly, as with lava flowing from a volcano, the rocks are commonly smooth, hard, and glassy. Obsidian is an example of such a rock. If, however, the cooling is very slow, the individual minerals separate out as large crystals. Granite is an igneous rock formed in this manner.

When igneous rock is exposed to the action of rivers, waves, and weather, it may be broken down. The products are then carried along and deposited in layers on the bottom of bodies of fresh or salt water. Over the years the particles become cemented together, with such things as organic debris and animal shells, to form *sedimentary* rock (Fig. 2-6). The accumulation of these particles over a long period of time frequently produces layer upon layer of deposited materials, often formed from particles of different size and composition. Thus deposits of sedimentary rocks can often be distinguished from igneous rocks by the presence of layers.

When igneous or sedimentary rocks undergo a marked change due to heat or pressure they are known as *metamorphic rocks* (Fig. 2-7). This process usually occurs deep within the earth.

WEATHERING

Any type of rock can become soil parent material after it has been broken down into small pieces by a process called *weathering*. After parent material is present, soil can form through the addition of dead organic matter, microorganisms, plants, and animals.

Weathering involves physical, chemical, and biological processes. These types of weathering are discussed separately here. Bear in mind, however, that they occur simultaneously in nature.

Slate

Schist

Gneiss

Fig. 2-7
Some metamorphic rocks. From what were these rocks formed?

Physical Weathering. During physical breakdown, rocks are split into smaller fragments with no change in chemical composition. This splitting can be brought about by several natural agencies including wind, water, ice, and temperature change.

Pieces of rock carried by wind or water wear down the surfaces over which they pass and are themselves broken down. Ice masses such as glaciers wear down the surfaces over which they move. (How do you suppose that ice is hard enough to actually break down the underlying rock?) Water freezes in rock cracks and crevices. As it freezes it expands. The resulting force of expansion causes rock breakdown. This phenomenon is known as *frost weathering*. The repeated heating and cooling of certain rocks, known as *insolation weathering*, leads to weaknesses in the rocks and causes splitting. Rocks which are composed of several minerals, each having a different coefficient of expansion, are most often disrupted in this way. (The coefficient of expansion of a material is the amount it expands or contracts when the temperature is increased or decreased 1 C°.)

Chemical Weathering. Chemical weathering involves actual chemical changes in the rock minerals. The newly formed substances are usually more soluble than the original minerals. Water is extremely important in the process of chemical weathering and may bring about hydrolysis, hydration, or simple solution.

During *hydrolysis*, water enters into a complex chemical reaction with the rock minerals. The result is the disintegration of minerals. Hydrolysis may be accomplished by water alone. However, it occurs more quickly in the presence of an acid. Water often contains carbonic acid and organic acids produced during the decomposition of organic matter. The acids provide hydrogen ions which speed up the reaction.

Hydration, a second form of chemical weathering, is the simple addition of water to minerals in rocks. This reaction frequently accompanies hydrolysis. When a mineral such as hematite (an oxide of iron) hydrates, it expands and softens, thereby contributing to the decomposition of the rock.

Water also acts to remove soluble products from the site of reaction, as well as to dissolve out easily soluble minerals that were originally present in the rocks. The accumulation of these soluble materials is reflected in the development of a particular soil. (See Section 2.3.)

Oxidation, the addition of oxygen, and *reduction*, the removal of oxygen, are two other chemical processes that change minerals into new substances and thereby aid in the decomposition of rocks.

Biological Weathering. This form of weathering is essentially a combination of physical and chemical weathering, brought about by the action of living organisms. The roots of plants can crack and break apart rocks. This occurs as the roots expand in rock crevices during growth (Fig. 2-8). Plant roots also produce carbon dioxide as they respire. This gas combines with water to form carbonic acid. Carbonic acid hastens the solution of certain minerals and thereby speeds up rock breakdown into soil parent materials.

Perhaps you have wondered how caves are formed. In the usual mechanism, carbonic acid attacks the limestone in rock formations. In so doing, it forms calcium and magnesium bicarbonate. These compounds are more soluble than any of the original rock. Thus they are carried away by the water flow, leaving large caverns in the rocks.

The decay of organic matter by plants and animals also produces carbon dioxide which weathers rocks by carbonation. Many burrowing animals, such as earthworms and woodchucks, expose soil particles to the air, thus assisting in their decomposition by chemical weathering.

Fig. 2-8
Is this form of biological weathering essentially chemical, physical, or both?

For Thought and Research

1 (a) Is the parent material in your area igneous, sedimentary, or metamorphic?

(b) Collect rock samples near your home and bring them to school. Compare them with those brought by other students. Are they all the same rock type? Why?

2 Describe the different pathways along which rainwater travels. Explain how this water weathers rock along these paths.

3 Weathering includes physical, chemical, and biological processes.

(a) Which of these require the presence of water? Why?

(b) Some of these processes act deep in the soil while others act only on the surface layers. List those that act on the deep soil layers and explain how they are accomplished.

4 Temperature changes can greatly enhance the weathering process. How? Be sure to include in your answer the effects of temperature on plants and plant growth. Do you think that physical weathering caused by temperature changes is a major factor in soil formation in your area?

5 (a) Why are rocks composed of minerals having different coefficients of expansion likely to undergo insolation weathering?

(b) Design and carry out an experiment to illustrate the role of temperature as a physical factor in soil formation.

6 (a) Find out what is meant by "the rock cycle."

(b) Find out how each of the rocks illustrated in Figures 2-5, 2-6, and 2-7 are formed.

(c) What role do you feel the rock cycle plays in soil formation?

7 Perform the field and laboratory studies in Sections 5.12-5.15.

Recommended Readings

1 *Geology Illustrated* by J. S. Shelton, W. H. Freeman, 1966. This well-illustrated book is an excellent source of information on rocks and geological phenomena. A brief account of weathering is included.

2 *Vegetation and Soils* by S. R. Eyre, Edward Arnold (Publishers), 1968. This book includes a thorough description of soil formation.

3 *Investigating the Earth*, Earth Science Curriculum Project, Houghton Mifflin, 1967. This book, written for secondary school students, clearly explains soil formation and development in Chapter 12.

4 Your provincial or state Department of Agriculture probably has published information on the soils in your region.

2.3 SOIL PROFILE DEVELOPMENT

Section 2.2 indicated that many processes act together on the earth's rocks to form the soil parent material. This parent material undergoes further natural changes to form soil. The initial step is brought about by the action of plants and animals.

Perhaps you have passed an area where a new road was being built and noticed a cut through a hillside. Looking closely at such a cut, one can often see layers of soil of different colors parallel to the surface of the ground (Fig. 2-9). These layers are called *horizons*. All of them together are referred to as the *soil profile*. The soil profile is commonly defined as the soil from the surface of the ground to the unchanged parent material beneath. The differentiation of soil into horizons is the result of the action of living organisms on the original parent material. The living organisms add organic matter to the soil and percolating water gradually carries the finer particles of soil material to lower levels. Since organisms and climate differ from place to place, soil horizons will also vary in composition and depth from place to place. They may also change with time in the same region.

Three main processes are involved in horizon development: accumulation of organic matter in the surface layers; eluviation (leaching of the profile); and illuviation (deposition of leached materials).

Organic Matter Accumulation. As you know, plants and animals are mainly responsible for the addition of organic matter to the soil. For this reason modifications of the soil vary according to a number of environmental factors, including parent material, temperature, and moisture. This is because these factors influence the development of the natural flora (plants) and fauna (animals).

On bare rock, for example, lichens and mosses are commonly the first plants to be established (Fig. 2-10). In order to extract required nutrients from the rocks, they secrete chemicals. This weathers the rocks and forms a layer of crumbled rock or mineral soil. As the lichens and mosses die, the humus content of the soil is built up and the rocky base is further weathered. Still more humus accumulates as wind and rain bring organic debris to the rocky area. This debris becomes trapped

Fig. 2-9
Close examination of a cut through a hillside often reveals one or more layers of soil on top of the bedrock.

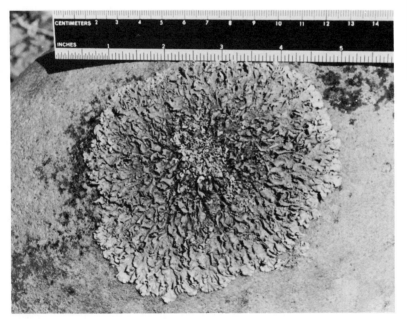

Fig. 2-10
Crustose lichens are usually the first plants to colonize bare rocks.

among the lichens and mosses. Eventually sufficient soil is built up to form a stable root anchorage. Ferns, grasses, and other herbaceous plants can now invade the area. Being larger and having a more rapid growth rate, these plants quickly build up humus as they grow, die, and decay.

While these changes are taking place, an increase in the number of microorganisms (bacteria, protozoans, and algae) also occurs. The organic debris formed as these plants die may accumulate in an undecomposed state on the surface. This gives rise to the uppermost defined soil horizon which is known as the A_{00} horizon (Fig. 2-11). Loose leaves and other unchanged organic debris can be easily recognized here. Just below this horizon is the A_0 horizon. Here decomposition of the organic debris is occurring. The A_0 horizon is usually divided into at least two layers, although they often merge imperceptibly into one another. The top layer is composed of slightly decomposed organic debris. Its original sources can still be easily recognized. This is termed the *fermentation* or *F layer*. Directly under this layer the physical structure of the organic matter changes due to more ex-

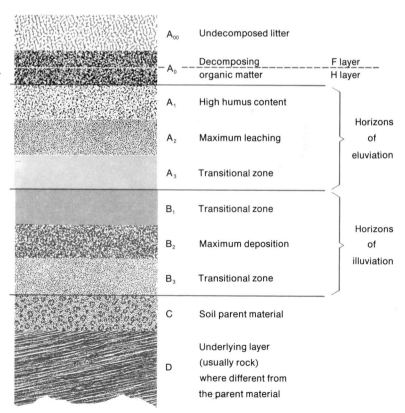

Fig. 2-11
Diagrammatic representation of a generalized soil profile.

tensive action by decomposers. This material constitutes the *humus* or *H layer*.

The same factors which control the formation of humus, apart from erosion by wind or water, control the development of the A_0 horizon. For example, reduction in the number of microorganisms favors the accumulation of surface litter and prevents the formation of an appreciable A_0 horizon. Similarly, the lack of soil oxygen, due to factors like a high water table, will also slow down the development of an A_0 horizon.

The soil fauna also play a significant role in the development of the A_0 horizon by mixing the surface litter with the mineral matter of the lower soil horizons. This occurs less often in acid soils because the organisms mainly responsible for this modification—earthworms and snails—prefer neutral and slightly alkaline soils.

Temperature also plays a part in the development of A_0 horizons. Low temperatures retard decomposition of organic matter. The consequence is an accumulation of surface litter. High temperatures have the opposite effect.

The nature of the organic matter is also a significant factor. Litter such as conifer needles contains large amounts of waxes and tannins. These decompose very slowly and, therefore, tend to accumulate. Deciduous litters, like the leaves of ash and maple trees, decompose more quickly since they are usually alkaline.

Eluviation. As you know, water can *percolate* (move downward) through the soil. When this water carries with it the soluble decomposition products of the A_0 horizon, it is responsible for a further stage in the development of the soil profile—*eluviation*. These percolating waters are acidic because some of the decomposition products of organic matter are acids. In addition, carbon dioxide formed during respiration by soil organisms reacts with water to form carbonic acid. The resulting acidic solution increases the solvent action of the water, speeding up the hydrolysis of primary minerals. (See Section 2.2.)

The continued action of such percolating water is known as *leaching*. Leaching is characterized by the removal of soluble material and the modifying of the residue to form the *A horizons* or *horizons of eluviation* (Fig. 2-12). The organic matter may also be affected by leaching. In fact, under conditions of intense leaching, the organic matter may be almost entirely washed from the upper horizons of the profile.

This is a simplified version of the leaching process. Section 2.4 elaborates more fully on this and the next topic, illuviation.

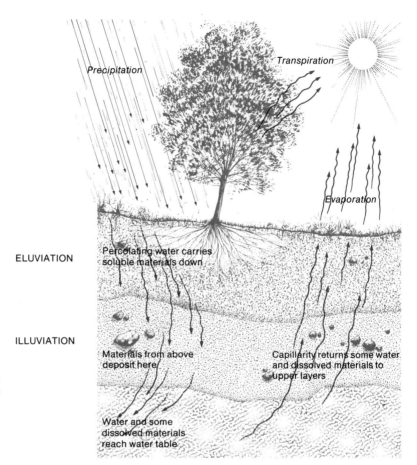

ELUVIATION

ILLUVIATION

Precipitation

Transpiration

Evaporation

Percolating water carries
soluble materials down

Materials from above
deposit here

Capillarity returns some water
and dissolved materials to
upper layers

Water and some
dissolved materials
reach water table

Fig. 2-12
The roles of leaching and
other phenomena in the
formation of a soil profile.

Illuviation. The soil profile does not necessarily lose the materials leached from the A horizons. These materials move down through the soil and are deposited in lower regions. There the rates of weathering and biological activity are much slower. This deposition of materials is called *illuviation*. The movement of materials from the A horizons may be checked or stopped by changes in the environment (such as decreased rainfall) or by an excessive accumulation of these constituents in the lower horizons. These lower horizons are known as the *B horizons*, or the *horizons of illuviation*. As with the horizons of eluviation, it is often possible to differentiate several horizons of illuviation. Each distinct layer is usually labeled separately, from the surface downward in each zone. Thus the horizons of eluviation are A_1, A_2, A_3, A_n, while those of illuviation are B_1, B_2, B_3, B_n. (See Figure 2-11 for the meanings of the subscripts.)

Beneath the B horizons is weathered parent material. This zone is called the *C horizon*. The next zone down is called the *D horizon* if it is of different origin from the parent material in the C horizon. (How can this be?)

All of these horizons are included in the soil profile shown in Figure 2-11. This diagram serves as a good review of this section and should be studied carefully. It represents a very generalized soil profile, and one that might develop under conditions of intense leaching in a temperate climate. Keep in mind that soils show wide variations of profile from place to place. These variations are not always easy to interpret in general terms.

For Thought and Research

1 (a) Suggest several natural ways by which an established soil profile may be altered.

(b) By what means can man slow down the effects of leaching and thus influence the development of the soil profile?

2 Suppose that instead of receiving abundant rainfall, the soil in a given area receives very little. In fact, the evaporation of moisture from the surface exceeds the precipitation. In which direction would you expect to find the soluble products of weathering moving? Which horizons would be enriched? With what? Would you find illuviation horizons in a soil profile from this area? Why?

3 (a) Why does less decomposition of organic matter occur in the F layer than in the H layer?

(b) The A_1 horizon is usually the zone of maximum biological activity. Why?

(c) Under what circumstances do you think the C horizon will consist of a different material than the D horizon?

4 (a) Suppose that the addition of a pesticide to a piece of land killed most of the earthworms in the soil. How would this affect profile development?

(b) Do you think that the development of a soil profile is necessary before the land is suitable for agricultural use? Why?

(c) Would you expect to find a soil profile in the soil of your school lawn? Why?

5 Perform the field and laboratory studies in Sections 5.20, 5.21, and 5.40.

Recommended Readings

1 *An Introduction to the Scientific Study of the Soil* by N. M. Cornber, Edward Arnold (Publishers), 1960. Read pp. 90-98 for an excellent discussion of soil development and soil profiles.

2 *Ecology and Field Biology* by R. L. Smith, Harper & Row, 1966. Soil profiles and soil formation are dealt with on pp. 243-252.

3 *Physical Geography* by A. N. Strahler, John Wiley & Sons, 1967. Read Chapters 16 and 17 for an adequate but brief discussion of soils.

4 *Principles of Physical Geography* by F. J. Monkhouse, University of London Press, 1968. This book contains a very good section on soil and soil types.

5 *The Living Soil* by J. R. Corbett, Martindale Press, 1969. This excellent book is written from a slightly different viewpoint.

2.4 SOIL CLASSIFICATION

By now you should know what soil is, how the parent material is weathered, and how the soil profile is developed. Review these concepts, if necessary, before you continue with the following study of the classification of soil.

One of the major problems in the classification of soils is the absence of a generally accepted terminology. Many schemes of classification have been proposed. These are clearly summarized in *Recommended Reading* 4. For our purposes we will subdivide all soils into three groups—zonal, intrazonal, and azonal. *Zonal soils* develop under conditions of good soil drainage and are the most widespread of soil types. Soils formed under conditions of very poor drainage (for example, bogs) are known as *intrazonal soils*. Soils which cannot be easily classified, either because they are very young, or because they have been unable to develop a soil profile, are known as *azonal soils*.

ZONAL SOILS

These soils have well-developed profiles due to the prolonged action of vegetation and climate. Zonal soils can be conveniently divided into two groups, the pedalfers and the pedocals. *Pedalfers* are soils in which the calcium carbonate content of the profile is *less than* that of the parent material. *Pedocals* have at least one soil horizon with a calcium carbonate content *greater than* that of the parent material.

In the pedalfers, the loss of calcium carbonate from the soil profile indicates that extensive leaching has taken place. Aluminum and iron are also carried down in the soil profile, but they tend to precipitate out and accumulate in the lower horizons. It is from the aluminum and iron accumulation that these soils get the name "ped-al-fers." Pedocals ("ped-o-cals") owe their name to the calcium carbonate that they accumulate. Roughly speaking, the soils of eastern North America are pedalfers since precipitation exceeds evaporation there. West of the 99th meridian in the United States (running from southern Texas to the eastern Dakotas), pedocals predominate because of the low annual precipitation. Variations do occur, of course, in mountain regions. Further, pedalfers are generally found in forest regions, while pedocals are in grassland and desert areas.

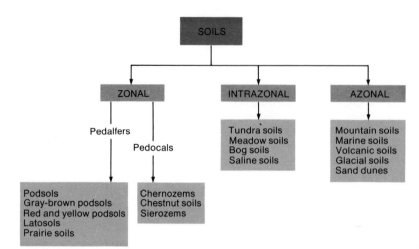

Fig. 2-13
A classification of soils.

There are several divisions of both pedalfer and pedocal soils. Figure 2-13 should help you to sort out these various soil divisions as you read the following descriptions.

Pedalfers

(a) Podsols. These soils are usually characterized by extensive leaching and, as a result, by a very distinct profile. The podsol profile has thick A_{00} and A_0 horizons. Immediately beneath them, the thin A_1 layer is very acidic and rich in humus. The acids from this layer increase the solvent powers of water. Thus extensive leaching occurs below this. The strong leaching that occurs in the A_2 horizon is known as *podsolization*. The materials (like aluminum and iron) leached from this horizon move downward to enrich the B horizons below. Because of the leaching, these soils typically have a white or gray A horizon over a brown B horizon. Virtually all minerals are leached from the A horizon, leaving it almost colorless. Minerals like calcium, nitrogen, and potassium leave the profile completely by leaching through to ground water level. Aluminum, iron, and organic material leach only as far as the B horizon where they are deposited, imparting the color to the soil (Fig. 2-14).

These soils are usually found in temperate, moist climates where precipitation exceeds evaporation. The percolation of the excess water into the soil causes podsolization. Since acid conditions aid podsolization, this process occurs best where the decaying vegetation produces an acidic humus. Coniferous forests (with the exception of pine) do this. Once established, such soils support mainly conifers. They are very low in minerals

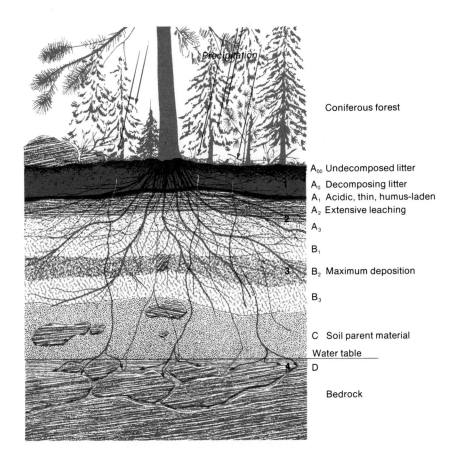

Coniferous forest

A_{00} Undecomposed litter
A_0 Decomposing litter
A_1 Acidic, thin, humus-laden
A_2 Extensive leaching

A_3

B_1

B_2 Maximum deposition

B_3

C Soil parent material

Water table

D

Bedrock

Precipitation

Fig. 2-14
Formation of a podsol soil, a representative of the pedalfer group. Describe what occurs at the depths indicated by the numerals 1 to 4.

and conifers need less of the basic minerals than do most other tree species. Podsols occur in the northeastern United States and from there extend north and northwest into Canada.

(b) Gray-Brown Podsols. These soils are also known as brown forest soils. They are characteristically found under deciduous forests. They are similar to podsols, but have a thinner A_0 horizon and less leaching of the A horizons. This occurs where rainfall is less severe, the soil is less permeable to water, and deciduous trees (such as beech, maple, and oak) return bases to the surface in the form of dead leaves. (The bases neutralize acids, slowing down leaching.) Thus the A_1 horizon is only slightly acidic. Materials are leached from the gray-brown A_2 layer and accumulate in the thick, dark brown B horizon. The gray-brown podsols are very fertile and form highly productive farmland when they are cleared of trees. Gray-brown podsols occupy a wide band through southern Ontario and the east-central United States.

(c) Red and Yellow Podsols. These soils show the same characteristic leaching of the A_2 horizon as the gray-brown podsols. The yellow soils have a grayish-yellow leached horizon over a yellow one; the red soils have a yellowish-brown leached horizon over a deep red one. They occur in warm climates where precipitation is abundant. The yellow soils occur in sandy coastal plain regions that are poorly drained; the red soils are found in well-drained upland regions. The humus content is low in both types of soil. The natural vegetation occurring on these soils, especially in the north, is the oak-pine forest. In southern areas, crops of cotton and tobacco are grown on these soils.

(d) Latosols. These soils form in areas of heavy rainfall and high temperatures, that is, in tropical and sub-tropical regions. Due to these conditions, the parent rock has been almost completely decomposed. Also, the low acidity produced by the decay of tropical litter dissolves the silica in the soil, allowing it to be leached by the heavy rains. Humus is almost entirely lacking due to the rapid action of decomposer organisms. Iron and aluminum compounds (usually oxides) tend to accumulate in the soil, giving it a distinctive reddish color throughout.

The removal of silica and the accumulation of iron and aluminum is known as *laterization*. The oxides which accumulate in layers are known as *laterites*. Valuable minerals are often deposited in laterites. Limonite (an oxide of iron) and bauxite (aluminum oxide) are two such minerals. They accumulate as the parent material wears away and as silica and other soluble constituents are leached out. Thus they are known as *residual ores*. Because extensive leaching has removed plant nutrients from all horizons but the thin organic layer, latosols are very infertile and will not support continued crop cultivation.

(e) Prairie Soils. Prairie soils are a transitional type between pedocal and pedalfer soils. They are located in a long strip through the midwest where precipitation is neither abundant nor sparse. They divide the forest soils of the east from the grassland soils of the west. The upper horizons of the profile are very dark brown in color with a gradation through to a lighter brown in the lower horizons. The typical vegetation on the prairie soils is tall grasses. Apparently these tall grasses can withstand the dryness of the soil between summer rains, while the deciduous forests bordering the prairies to the east cannot.

Pedocals

The second group division of zonal soils, the pedocals, develop under conditions of slight rainfall. Thus they are incompletely leached soils, and calcium carbonate (formed by carbonation in surface layers) deposits in the B horizon. (Recall that calcium is completely leached from pedalfer soils.) Also, the ground water, carrying soluble minerals with it, moves upward due to the combined effects of the rapid evaporation at the surface and capillary action. This adds even more calcium carbonate to the upper layers of the soil. Several of the more common pedocals are discussed here.

(a) **Chernozems**. The chernozems or black soils are developed typically in grassland areas. Due to the cold winters, and hot summers, with periods of drought and strong evaporation, weathering occurs, yet excessive leaching is not present. This forms a very characteristic profile (Fig. 2-15). The A horizon is black, high in organic matter, and very thick (about 1 meter in some cases). Moving down toward the B horizon, the soil becomes lighter in color, until in the B horizon the soil is brown or yellowish-brown. The C horizon is very light in color and dry (because of a low water table). Although the A_2 horizon is not leached, soluble minerals carried from the surface downward through the A horizon do accumulate in the B horizon. Capillary action brings minerals back up to the A horizons. Thus the whole profile is saturated with calcium carbonate and other minerals.

Fig. 2-15
Formation of a chernozem soil, a representative of the pedocal group. Describe what occurs at the depths indicated by the numerals 1 to 4.

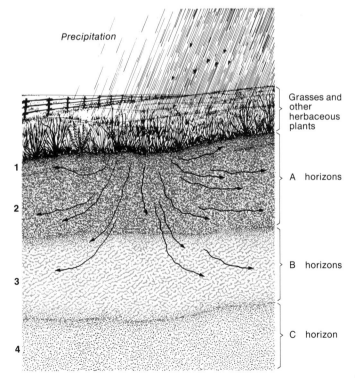

Precipitation

Grasses and other herbaceous plants

A horizons

B horizons

C horizon

Since nutrients are not leached away to any great extent, the soil remains highly fertile. Forests cannot exist on chernozem soils, mainly because they cannot withstand the periods of drought. However, many grain crops (barley, oats, rye, and wheat), which are drought-resistant and tolerant of excesses of minerals, do well on these soils. Chernozems occur in a wide band from Canada to Mexico, immediately to the west of the prairie soils.

(b) Chestnut or Brown Soils. These soils develop in semi-arid regions. Because the soil profile contains less humus than the chernozem, it is lighter in color, often grading into lime accumulations. Otherwise, the soil profile is very similar to that of the chernozem soils. These soils usually develop under a growth of short grasses and can be fertile when moisture in the form of irrigation or rainfall is available. They border on the west of the chernozems and occupy the region commonly called the Great Plains.

(c) Sierozems. These soils are found in desert regions. Thus they are very low in organic matter due to the sparseness of the vegetation. This accounts for their pale gray color. The soil profile contains horizons, but these are often difficult to see. Due to the prolonged dry periods, when evaporation from the surface permits a rising of the ground water, a lime crust of calcium carbonate and calcium sulfate forms at the surface. This accumulation of salts in the surface horizons also partially accounts for the light color of these soils. Extensive irrigation is required before crops can be cultivated on these soils.

Figure 2-16 shows the relationship between climate (temperature and precipitation) and soil type for the zonal soils discussed here. Figure 2-17 shows their locations. Study these figures carefully to review zonal soils.

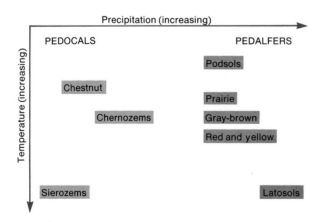

Fig. 2-16
Relationship between climate and soil type for the major zonal soils.

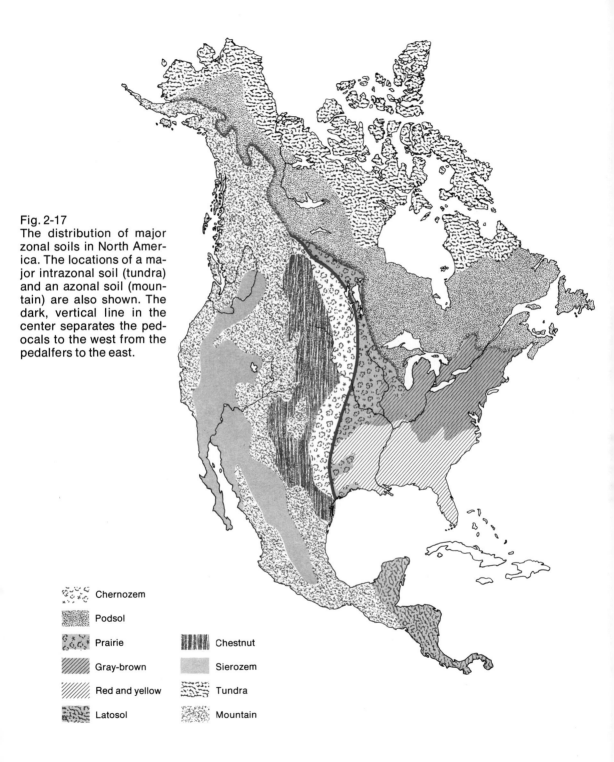

Fig. 2-17
The distribution of major zonal soils in North America. The locations of a major intrazonal soil (tundra) and an azonal soil (mountain) are also shown. The dark, vertical line in the center separates the pedocals to the west from the pedalfers to the east.

Chernozem

Podsol

Prairie   Chestnut

Gray-brown   Sierozem

Red and yellow   Tundra

Latosol   Mountain

INTRAZONAL SOILS

Intrazonal soils develop in areas where the drainage is very poor. There are four main types of intrazonal soils—tundra soils, meadow soils, bog soils, and saline soils, each determined by unique climatic conditions (Fig. 2-18).

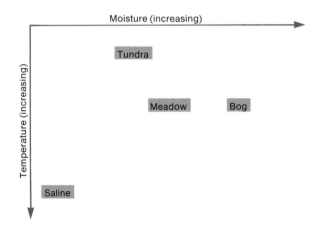

Fig. 2-18
Relationship between climate and soil type for the major intrazonal soils.

(a) **Tundra Soils**. These soils are formed in areas where the temperature and precipitation are low and where the growing season is very short. A large amount of humus is found in tundra soils because the rate of decomposition of organic matter is slow in such a harsh climate. Distinctive soil profiles do not form. Instead, the humus usually lies on a permanently frozen layer.

(b) **Meadow Soils**. Meadow soils form on stream flood plains where drainage is poor due to flooding part of the year. This flooding is accompanied by silt and mud deposition. Therefore a thick, dark, humus-rich surface horizon is developed. Lower down in the soil profile, the results of waterlogging are clearly visible.

(c) **Bog Soils**. These soils form in marshy areas where the water table is very high, saturating the soil during most of the year. This severely retards plant decay. Where plant decay does occur, the acids produced are not readily neutralized. Thus the humus tends to be highly acidic. Because of the high acid content and marshy condition, this type of soil is not very useful for agricultural purposes.

(d) **Saline Soils**. Areas where evaporation from the soil greatly exceeds precipitation have saline soils. Soluble salts of sodium, potassium, magnesium, and calcium are accumulated in the upper horizons of the soil profile. These salts have been carried upward by capillarity and have remained when the water evaporated. Vegetation is sparse on these soils. However, crops can be produced if the salt concentration is lowered and drainage is provided.

AZONAL SOILS

Azonal soils have poorly developed soil profiles. Some are young soils and have not had sufficient time to undergo weathering and other soil-forming processes. Others occur in areas where new soil material is being continually deposited.

Mountain soils occur on rocky and unstable slopes where the developing horizons are mixed long before they can achieve maturity and thus be classified as zonal soils. Marine soils in the form of mud and sand banks along low-lying coasts have basically the same situation. Volcanic soils, where recently deposited ash, lava, and pumice are common, are able to effectively offset the process of leaching. Other azonal soil types include glacial soils and wind-blown soils such as sand dunes.

For Thought and Research

1 (a) Bog soils are not good agricultural soils. Why?

(b) List and explain three ways by which bog soils could be improved to produce crops.

2 (a) What are laterites?

(b) Sketch a typical lateritic soil profile showing the minerals present in each horizon. Where are these soil types generally found?

(c) Would you consider this type of soil to be highly productive? Why?

3 Podsol soils have a very characteristic soil profile.

(a) Describe the climatic conditions necessary for the development of this profile.

(b) What natural vegetation would you expect to find growing on podsol soils?

(c) Are podsol soils generally considered to have a high agricultural productivity? How could their productivity be improved?

4 Tundra soils, being intrazonal, differ considerably from the podsol soils.

(a) Compare the soil profiles of these two soil types in detail. Drawings of their soil profiles may make your comparison easier to follow.

(b) Because tundra soils develop under conditions of excessive cold, the composition of the parent material of the soil is affected. How?

5 Soils of dry climates differ fundamentally from soils of wet climates.

(a) Draw and label two soil profiles, one to show the effects of leaching and the other to show the effects of capillary action on a soil.

(b) Explain in detail the mechanisms involved in the development of the soil profile in each case. Be sure to mention precipitation, evaporation, and the initial and final positions of the soluble materials.

(c) Two major soil types are recognized on this basis. What are the names of these soil types and where in North America would you expect to find each one?

6 Imagine that you have been given your choice of 100 acres of farm land anywhere in North America. In order to become a successful farmer you must choose the very best soil for your farm.

(a) Would you choose a zonal, intrazonal, or azonal soil type? Why?

(b) Where would you choose to have your 100 acres of farmland? Discuss in detail the reasons for your choice.

7 Consider the soil formation controls and agents which exist in the area where you live. Describe the soil profile which you would expect to be developed. What kind of soil is it? Is this soil type the same as the zonal type in your area? If not, why not?

8 (a) Account for the deep, organic-laden A horizon of chernozem soils.

(b) Chestnut soils in the south are lighter in color (contain less organic matter) than those in the north. Why?

(c) Why is the humus content low in red and yellow podsol soils? What evidence is there that these soils undergo some laterization?

9 Three major processes—podsolization, laterization, and calcification—are involved in the formation of soil profiles that reflect local climatic conditions. Distinguish between these three processes by indicating the climatic conditions under which each takes place and by describing a representative soil profile formed by each process.

10 Perform the field and laboratory study outlined in Section 5.41.

Recommended Readings

These books all contain information on soil classification.

1 *Principles of Physical Geography* by F. J. Monkhouse, University of London Press, 1968.

2 *Ecology and Field Biology* by R. L. Smith, Harper & Row, 1966.

3 *The Living Soil* by J. R. Corbett, Martindale Press, 1969.

4 *The Nature and Properties of Soils* by H. O. Buckman and N. C. Brady, Macmillan, 1969.

5 *Vegetation and Soils* by S. R. Eyre, Edward Arnold (Publishers), 1968.

6 *The Study of Plant Communities* by H. J. Oosting, W. H. Freeman, 1956.

Life in the Soil: Macrofauna

3

Beneath the surface of the ground lives a complex community of organisms which is dependent upon dead organisms from the community above for its energy. In this community the organisms vary in size from minute viruses, visible only with an electron microscope, to large vertebrates like gophers. But, large or small, each one of these organisms plays an important role in the intricate web of life of the soil ecosystem.

Our investigation of life in the soil begins with the macrofauna ("large animals"), those animals that can be seen without a microscope. This Unit describes many of the common macrofauna and the niches that they occupy in the soil ecosystem.

3.1   EARTHWORMS AND OTHER ANNELIDS

Any good fisherman can tell you a bit about earthworms. A little earthworm ecology, after all, can come in handy if you are planning a day at the lake or stream and wish to obtain a can of juicy fish bait. Here are some questions you may have already asked on some occasion: Where are the greatest numbers of earthworms to be found? When are they most easily caught, and how? Once caught, how can they be kept alive? These are

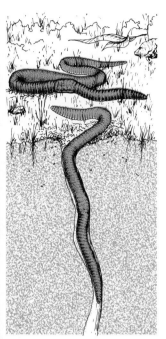

Fig. 3-1
The common earthworm,
Lumbricus terrestris.

practical questions when asked by the fisherman. But they might also be asked by the soil ecologist wishing to find out what earthworms require in order to survive.

Where, in fact, can one find the greatest number of earthworms? The large "common" earthworm, *Lumbricus terrestris*, is the earthworm with which most people are familiar (Fig. 3-1). These animals are abundant in many areas. They survive best in moist, loosely packed soil containing lots of organic matter and covered with a fair amount of litter, such as dead leaves or grass. Why are these conditions so favorable? Since earthworms do not have a hard outer shell like insects, they lose moisture if the soil in which they live is very dry. Even the best of soils may get fairly dry between rainfalls. In such cases worms can lose up to 70% of their water content, but they immediately regain this loss when water is again available.

Ideally, a permanently moist soil would cause no hardship to the worms since no water loss would occur. Why then are loosely packed soils best? In the first place, clay soils, which are normally quite compact, contain very little air and thus provide only a small amount of vital oxygen. Soils that are loose (many spaces between the soil particles) can contain plenty of air and oxygen. In addition, earthworms can burrow much more easily through loose soil. They easily force their way through the crevices and air spaces, pushing soil to the side as they form their tunnels. When necessary, the pathway can be extended by swallowing the soil. The swallowed material may or may not contain food items. The soil travels through the body of the animal and is then plastered with mucus into the small air pockets and spaces along the sides of the burrow. This results in a continuous tunnel with a distinct lining of digested soil. If there are few air spaces, tunneling is much more difficult. Without places to deposit the waste, the worm must take it to the surface and pile it up outside the entrance to the burrow. This is often the case in compact clay soils. You have probably seen small mounds of so-called "casting" on the surface of hard soil. Earthworms often obtain food while tunneling, which is another reason why they are found in soils containing a lot of organic matter.

Earthworms also need their burrowing abilities with the approach of winter. At this time of year, they must retreat to depths below the frost line to avoid the killing cold. Deep vertical burrows are also formed to avoid drying out when rains fail to keep the upper layers moist. In this case, the worms may descend as far as the permanently moist soils at the top of the water table, possibly 3 or 4 meters below the surface.

Where is the ideal place to find earthworms? Look for soil containing a fair amount of humus and a good depth of surface litter. Not many people realize that earthworms do much of their feeding right at the surface. A good fisherman knows that, during the night, especially when the upper soil layer is moist from a recent rain or when the grass is bathed in dew, the large earthworms cautiously glide part way out of their burrows, glistening in the light of the stars. Their interest is not in stargazing, since their eyesight is extremely poor. Nor is their interest always in mating, which incidentally often does occur on the surface. The so-called "dew" worm is in search of food. With their "tails" firmly anchored in the ends of the burrows, they delicately probe the surface world in search of dead plant material. If successful, a leaf or a few dead blades of grass may be found and pulled down into the entrance of the burrow. Those plant fragments that make it into the burrow are coated with a slime material which helps to decompose them to the point where they are easily swallowed. The wise worm, if there is such a thing, completes its surface feeding before sunrise. On the surface the worm is quite vulnerable. As the saying goes, "the early bird gets the worm!"

Soil miner, earth excavator, or master tunneler—any one of these terms could be used to describe the role of the earthworm. Although the life of the lowly worms must be extremely dull, their importance in the soil cannot be ignored. This one group of animals does more to alter the nature of the soil physically than all other soil animals combined.

Realistically, what do an earthworm's efforts achieve? Most important are the thousands of miniature mine shafts that these worms create. They give the soil the chance to breathe. Root cells need oxygen as much as leaf cells do. Vertical shafts, deep into the soil, make it possible for the soil air to circulate and get new supplies of oxygen. In the same manner, water which might otherwise run off the surface of the soil into streams quickly enters the soil via the same shafts to reach roots far below.

Earthworms, as they dig into deeper levels, pass soil that is rich in minerals up to higher levels, within the reach of the surface roots. The top surface layers might become exhausted of certain minerals needed by plants if worms ever went on strike. Possibly more important is the fact that dead leaves and other surface remains are quickly returned to a ground-up mulch, and mixed with inorganic soil particles. Organic waste left by worms is then attacked by soil decomposers and returned to a state where plants can again absorb essential nutrients from it.

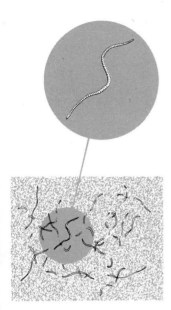

Fig. 3-2
An enchytraeid, or "white worm."

So far, this discussion has concerned only those worms with which most people are already familiar, the earthworms. This group is well known because its members are big and readily observed. A great many organisms have features very similar to earthworms and have been included with them in the single category called annelids (segmented, worm-like animals).

Another annelid group living in soil is a family of tiny worm species 5 to 15 mm long, called enchytraeids (Fig. 3-2). Because of their whitish color, they are commonly called "white worms." Unlike earthworms, enchytraeids are confined to the top, organically-rich soil layers penetrated by plant roots. Due to their small size, they show little ability at "eating" their way through the soil. Instead they wriggle their way along through air spaces and crevices. Unable to burrow deep into the ground to escape surface soils that might dry out, they are generally restricted to areas that remain permanently wet. Under such conditions, feeding is a rather simple process. Enchytraeids simply secrete a strongly basic (alkaline) solution. With this liquid they coat the moist plant remains in the upper humus. When the plant tissues have dissolved, they can easily be eaten and the waste further mixed into the humus.

White worms are ideal food items for aquarium fish, leeches, small amphibians, and crayfish. Should you require such food in your classroom, a technique is outlined in Section 5.33 for raising white worms in the laboratory.

Try the field and laboratory studies outlined in Sections 5.24, 5.25, 5.26, and 5.33.

3.2 ROUNDWORMS

Biologists say that one of the trademarks of success in the animal world is the ability to survive in a large number of different habitats. If this is so, the roundworms, more properly called *nematodes*, are truly successful. Members of this group occupy nearly every major habitat in the world. There are more than 10,000 known nematode species and new discoveries are made frequently. A number of these species can be found throughout the world in such widely separated locations as Hawaii, Antarctica, Canada, and Britain. Many are successful as free-living forms, while others exhibit varying degrees of parasitism in such diverse hosts as marine invertebrates, plants, and animals (including man).

About 2,000 species of nematodes inhabit the soil and about half of these are known to be true soil species. In some cases, as many as 30 million nematodes live in a cubic meter of soil, but more conservative estimates place the figure between 5 and 12 million in most soils. This is still sufficient to earn the nematodes the distinction of being the second most abundant form of animal life in the soil community. Only the microscopic protozoa exceed them in abundance. In fact, if you follow the classification system used by many biologists, nematodes rank first among animals since protozoa are not classed as animals in that system.

The name nematode means "thread-like." This may be the best way to describe these tiny, unsegmented worms, for they do indeed resemble small pieces of fine thread. They are extremely thin and range in length from 0.5 to 1.5 mm. Under a microscope their complex structure shows up well (Fig. 3-3). They have a definite head region equipped with bristles that serve as sense organs. The mouth opens into the *buccal cavity* which may possess "teeth" or some type of grinding surfaces, depending upon the species. Next in line is the muscular esophagus which forces food directly into the intestine and eventually out through the anus. Many nematodes possess an excretory pore on the ventral surface. Through this opening they secrete a slime that aids in locomotion.

Reproduction in the nematodes may be accomplished in a number of ways. In some species there are males and females which reproduce sexually. After copulation the female lays her eggs. One female of the genus *Cephalobus* laid 209 eggs in 20 days. This was the equivalent of five times her body weight. After a period of time, the eggs hatch and larval forms emerge. Originally these are about 0.5 mm long but they grow quickly. As they grow, they shed their skin. Usually they molt four times before becoming sexually mature in about 20 to 30 days. This time period varies considerably though, and many forms take much less time to mature. In some cases, the skin remains attached to the larva, forming a capsule which protects the larva from adverse conditions.

In many species, males have not been found. In such cases the nematodes are either *hermaphrodites* or they reproduce *parthenogenetically*. Hermaphrodite individuals possess complete sets of male and female reproductive organs. Usually, however, these are not both functional at the same time. The sperm develop first and then the eggs ripen. Parthenogenetic reproduction involves the normal development of an organism from eggs which have not been fertilized. In either case the females can carry on their existence and give birth to the next generation without the aid of males.

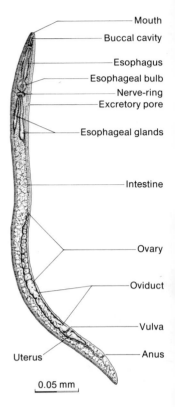

Fig. 3-3
Microscopic view of a typical soil nematode.

Fig. 3-4
Typical mouth structure and anterior digestive tract of a bacterial-feeding nematode. Note the muscular esophagus.

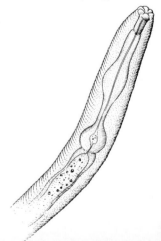

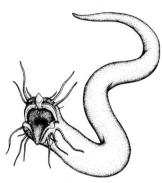

Fig. 3-5
A predatory nematode.

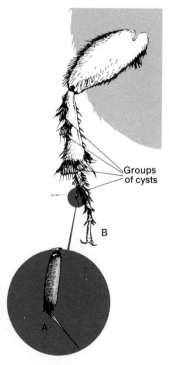

Groups of cysts

B

A

Fig. 3-6
Rhabditis coarctata. (A) a larval cyst; (B) cysts attached to the leg of a dung beetle.

Nematodes display a wide variety of living habits. Many of them are free-living forms which spend their time wriggling through the layers of moisture that coat soil particles. The smaller free-living forms feed on bacteria, small algae, and actinomycetes. Their digestive systems are well-adapted to this mode of life. They possess "teeth" in the buccal cavity or a specially adapted esophagus for breaking up the small organisms that they ingest (Fig. 3-4). They often take in tiny particles of grit that also aid in this process.

Less prominent, but still of importance, are the larger predatory forms (Fig. 3-5). These fierce creatures possess strong buccal teeth which enable them to grasp and devour smaller nematodes and other worm-like organisms. One ravenous predator consumed 83 nematode larvae in a single day. The same individual ate 1,322 larvae over a period of 12 weeks. This prompted researchers to explore the possibility of using these predators as a means of biological pest control but, to date, these methods have not proven very successful.

At one time scientists thought that certain nematodes were saprophytic, but recent studies tend to discredit this belief. It is true that a tremendous increase in the number of nematodes occurs when fresh organic matter is added to the soil. This, however, is now accredited to forms which feed on the microbial life that accompanies such an addition, rather than on the decaying organic matter itself. One such species is *Rhabditis coarctata.* The larval cyst of this nematode will even attach itself to the leg of a dung beetle (Fig. 3-6) so that it can be transported to a new food supply.

A large number of soil nematodes are parasites. Some of those which infect animals use the soil merely as a stopover between hosts, and are not true members of the soil community. Those that parasitize plants are far more important because of the damage they can do to agricultural crops. Members of this group usually spend their active life in some part of a plant. Their eggs are enclosed in a cyst which can remain inert in the soil for years. The eggs hatch when they are stimulated by a secretion from the proper host plant. The larvae then make their way to the plant to begin the cycle again.

One example of this type of nematode is *Heterodera rostochiensis.* This species, which attacks potatoes and tomatoes, has been responsible for great commercial losses. The eggs of this nematode are enclosed in a cyst which lies dormant in the soil for years. A cyst 0.5 to 1.0 mm in diameter may contain as many as 600 eggs. Each year a few eggs hatch but the larvae die. When the remaining eggs are stimulated by a secretion from the root of a suitable plant, they hatch and hundreds of larvae are released into the soil. These tiny larvae, 0.5 mm in length, follow the chemical trail to the root. They enter the plant by forcing their buccal stylets into the root (Fig. 3-7). Once inside, they cause certain root cells to swell and block the flow of sap in the root. The nematodes then grow and develop. The female becomes swollen with eggs. As she gets bigger her body is forced to the outside of the root while her head remains embedded in root tissue. The male leaves the root and fertilizes the swollen female. With his job completed, the male dies. The female's body becomes tough and leathery and she, too, soon dies. All that remains is a bag of eggs which falls off the plant when it is harvested. The plants are able to tolerate some degree of infection without any ill-effects. Some plants can support thousands of these freeloaders and not show any signs of damage. They simply send out more roots to compensate for the blockage. However, if the infection is extensive enough, there is little growth. Most of the leaves die and the plants are stunted in growth.

Some potato plants have shown varying degrees of resistance to these pests. At the present time, researchers are not sure why this is so. It may be due to a toxin produced by the plant that hinders the nematode, or it may be that the plant does not supply the nematode with all the nutrients it requires. Whatever the reason, this could be a valuable clue to a future method for control of these parasites.

The effect of this type of parasite is not always detrimental. In many instances the root-knot nematode, *Meloidogyne*, does cause a great deal of damage (Fig. 3-8). But in North Africa it actually benefits its host. An infection of this nematode increases the host plant's water-holding capacity. This results in increased drought resistance, an obvious advantage in dry regions.

Regardless of the source of their food or the manner in which they obtain it, nematodes feed primarily on cell protoplasm. They lack the ability to digest other substances. In fact, when a predatory form eats a small nematode, the victim will pass right through the predator's digestive system unless its cuticle is ruptured.

In general, soil nematodes tend to be located in the top 5 centimeters of the soil. They are rarely encountered below the 10 centimeter mark. In the upper layers the smaller free-living forms find adequate supplies of microorganisms on which to feed. The parasitic forms are attracted to the same areas because of the extensive root growth there. Finally, these forms attract

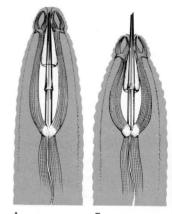

Fig. 3-7
Mouth structure of a plant parasitic nematode, showing the stylet in (A) the retracted, and (B) the extended position.

A   B

the larger predatory nematodes. Many other environmental factors influence this distribution of nematodes, but scientists have not yet been able to gain a clear understanding of just what they are and how they operate. Because of their great abundance, it is logical to assume that the nematodes are important members of the soil community.

Very few animals possess the chemical machinery necessary to decompose complex plant tissues. The nematodes are no exception. Any effects they may have on the decomposition processes in soil are indirect. It is possible that by feeding heavily on microorganisms associated with decay, the nematodes could influence these processes. This is an important area, but it needs a great deal of further study.

The predatory nematodes are noted for their tremendous appetites. They prey on a variety of microfauna including protozoa, other nematodes, and small enchytraeids. Their role as agents in the control of these organisms should not be overlooked.

It is ironic that the second most abundant animal in the soil community should be shrouded in such mystery. Unfortunately, many researchers have neglected this creature. To date, knowledge about nematodes has been based largely on the agriculturally important species. Only recently have researchers become interested in studying nematodes for their own sake. Perhaps the near future will yield more valuable information about these important soil animals.

You can collect and study nematodes by performing the field and laboratory study outlined in Section 5.34.

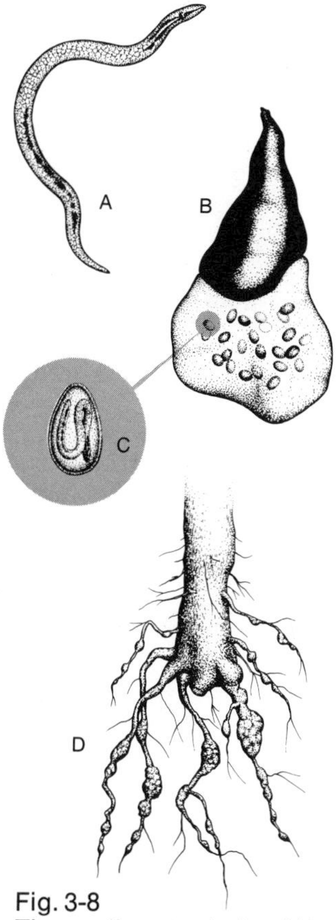

3.3   SNAILS AND SLUGS

Most major groups of animals have members that live in the soil. The Mollusca is no exception. While it is true that most molluscs are aquatic, certain species, particularly among the snails and slugs, are members of the soil community.

Terrestrial snails and slugs resemble each other in a number of ways. Even the most inexperienced observer should be able to note many similarities. Both have a soft upper body resting on a single muscular "foot" that runs the entire length of the organism. At the front end both have two long tentacles which are used in a way similar to that in which a blind man uses his white cane. Slightly behind these are two other stalks which support a pair of eyes (Fig. 3-9). Thanks to these specialized sensory organs and to concentrations of nerves in various parts

Fig. 3-8
The soil nematode *Meloidogyne*. (A) the larval stage; (B) a swollen female with her egg sac; (C) an egg with developing larva inside; (D) root-knot disease on a tomato plant due to heavy *Meloidogyne* infestation.

of their skin, slugs and snails have a sensitivity to light and sound, in addition to a well-developed sense of smell.

The only major difference in appearance between the two is that snails have a shell and slugs do not. Even this is not completely true, for some species of slugs still possess remnants of a shell underneath their skins. And, in certain soil-dwelling snails, the shell may be so reduced that, at a glance, the creature might be mistaken for a slug.

The snail's shell is really his home. It affords the snail many of the same benefits that our houses afford us. It offers protection from adverse environmental conditions. When winter approaches, the snail withdraws into its shell, seals the open end with fast-drying mucus, and awaits the return of warmer weather. The shell is also the snail's final defense against predators. But, contrary to some beliefs, the snail cannot leave its shell behind. It is an integral part of the organism.

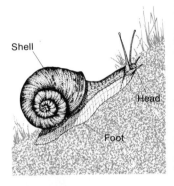

Fig. 3-9
A common soil mollusc, the garden snail.

What about the unfortunate slug? Is it devoid of protection and a home? On the contrary! Slugs are covered with a slimy secretion which offers some protection. Without the bulky shells, they are able to take refuge under litter and stones and in small cracks in the soil. In adverse conditions most slugs secrete a thick coating of slime about themselves and hibernate in the soil or under layers of litter.

Both the snail and the slug are hermaphrodites. That is, each has a complete set of both male and female reproductive organs. However, because these systems are not functional at the same time, individuals must mate with one another in order to reproduce. Depending on the species concerned, mating can take place throughout the year. But it often reaches a peak during late summer and early fall. The eggs, which are rather small, are usually laid in bunches, but some species deposit them singly throughout the soil. The eggs hatch in about three to four weeks and the young mature very quickly.

Snails and slugs tend to be nocturnal in their habits. They are usually active from two hours after sunset until two hours before sunrise, but some species overstep this general limit. For example, the common garden slug often begins his nightly expeditions before sunset and continues until after sunrise the next morning. During the daylight hours, snails and slugs tend to remain out of sight, hidden under leaf litter or buried in the soil. This rhythm allows them to be active during the warm, moist nights while escaping the heat of the day.

During their foraging expeditions, these creatures seem to glide along effortlessly on their elongated feet. A gland near the front edge of the foot secretes a slime that lubricates the foot and prevents its delicate surface from being torn to shreds by the underlying surface. Wave-like muscle contractions traveling from the back of the foot to the front move the snail or slug along at speeds of up to seven feet in five hours. It is usually easy to see where these animals have ventured. Wherever they go, they leave a ribbon of slime which dries and glistens in the sunlight.

Unlike many other soil inhabitants, snails and slugs cannot excavate their own tunnels through the soil. Therefore they are confined to the spaces between layers of litter and to loosely packed soils.

As a group, the terrestrial snails and slugs exhibit a wide variety of feeding habits. In fact, they eat nearly every type of edible material available and they all feed in much the same manner. Usually they roam about in search of a suitable food supply. Once they have found one, they ingest as much food as possible by scraping at it with a rasp-like tongue more properly known as a *radula*. This is really a hard ribbon of chitin with many recurved teeth. It lies in the bottom of the animal's mouth. The radula of each species has a characteristic structure which can often be used as an aid in identification. Once the food is ingested, it is stored in the crop until the snail or slug returns home where it can digest the food at its leisure.

Many snails and slugs are strict vegetarians. They ingest a wide range of living plant materials and, as gardeners will confirm, they can do considerable damage to important agricultural crops. Members of this group often feed on vegetation that hangs close to the soil and on the tops of young plants as they push up through the surface of the soil. However, *Milax budapestensis*, a common slug (Fig. 3-10), feeds below the surface where it attacks potato tubers. It is one of the few slugs that can penetrate the tough skin of the potato. In one instance, a crop of potatoes was infested with these slugs and all that remained of the entire crop were empty potato skins. Other members of this group feed on algae, lichens, and fungi.

Some species are saprophytes. Their primary sources of food are decaying leaves and woody tissues. Some snails and slugs secrete their own cellulose-digesting enzymes and are not dependent on bacteria to digest the cellulose for them as are many other so-called wood-eating organisms.

A few snails and slugs are carnivorous. They may feed on other snails and slugs and on small worms. If the prey is small, it is often grasped with the radula and swallowed in one piece. With larger prey, the snail or slug usually devours its victim piece by piece. One rather large slug, *Testacella*, feeds

Fig. 3-10
Milax, a common slug, feeds on potato tubers under the soil.

primarily on earthworms. It is able to elongate its body and pursue the earthworm into its burrow. The head-on battle that follows is more like a tug-of-war than a fight to the death, but the persistent slug rarely releases its grip.

There is an interesting oddity related to the feeding habits of these terrestrial molluscs that still puzzles scientists. An animal has elements in its body that are obtained from the food it eats. The blood of these molluscs is rich in a bluish-colored copper compound called haemocyanin. But the plants eaten by these creatures contain only the faintest traces of copper, and the soils in which they live are often quite deficient in copper. As of yet, researchers have not been able to determine where the snails and slugs obtain the copper.

The distribution and abundance of these creatures vary. The presence or absence of moisture, shelter, and food seem to be the main controlling factors. Snails and slugs are most abundant in the surface layers of the soil and in the overlying litter (Fig. 3-11). Here the substrate is generally porous and the organisms in question can move about without much difficulty. Food, moisture, and shelter usually abound here as well. One notable exception is the snail *Caecilioides acicula* which lives exclusively in the deeper regions of the soil (Fig. 3-12). It is usually found at depths near 40 centimeters where it crawls through existing pores in the soil and feeds on fungi.

The presence of calcium is another important factor in the distribution of snails. Their shells are composed mainly of calcium compounds, so it is only logical that they would be scarce in regions where the soil is lacking in calcium.

Estimates of the size of molluscan populations vary considerably. The activity of these creatures is so variable and dependent on such a variety of environmental factors that researchers find it difficult to arrive at accurate estimates. Some figures range from 7,500 snails per acre to over 600,000 per acre, depending on the type of vegetation and the soil conditions. Early estimates were based on the number of organisms captured during a timed walk through a garden. One researcher captured 570 slugs in a 30 minute search and 517 in a similar foray one-half hour later. However, these estimates are extremely crude and scientists are still trying to find a better method to determine population size.

By virtue of their feeding habits, many terrestrial molluscs play an important part in the decomposition of the organic litter layer. They ingest a great deal of this organic material and break it down both physically and chemically. The fact that they can produce their own cellulose-digesting enzymes

Fig. 3-11
Vollonia is a common inhabitant of the litter layer.

Fig. 3-12
Caecilioides acicula is a blind, colorless snail that lives in the deeper layers of the soil.

makes them of great importance in the decay of woody tissues which are resistant to many other organisms. Unfortunately, their role in such decomposition processes is very dependent on soil conditions. As mentioned earlier, snails and slugs depend on pre-existing channels for movement through the litter. If the litter layer of the soil is compressed, as is the case in certain areas, their ability to function is curtailed to a great extent.

Snails and slugs produce a great deal of mucus, both in their gut and on the surfaces of their bodies. This sticky slime tends to bind soil particles together and is an important factor in improving soil structure.

Certain groups of snails play a major role in the initial colonization of rocky areas. These types feed primarily on lichens which grow over the rocks. Some of the lichens grow several millimeters into the rocks and send out horizontal branches just under the surface of the rock. These snails are able to rasp away the overlying rock and devour the delicate lichens. Their droppings are rich in nutrients and provide ideal substrates for further colonization of such rocky areas by more advanced plant life.

These strange, sluggish creatures are valuable members of the soil community and the fact that they are so widespread serves to enhance their importance. But we must not overlook the fact that certain species are capable of doing a great deal of damage to vegetation.

3.4 ARTHROPODS

If numbers mean anything, arthropods are, without doubt, the most successful group of animals. Over 750,000 species belong to the phylum Arthropoda. Of these, almost 700,000 are insects. All told, arthropods constitute over 90% of the animal kingdom, and they inhabit all corners of the world.

Arthropod means, literally, "jointed feet." All members of this group of animals have jointed appendages, not just jointed feet. In addition they have segmented bodies, although the segments are often fused. Both the bodies and appendages of arthropods are covered with an exoskeleton (external skeleton) composed of *chitin*.

Most common among the arthropods are insects, spiders, centipedes, millipedes, crayfish, lobsters, and crabs. In this section, we will examine the common arthropods that inhabit the soil. As you read about the various soil arthropods, perform the appropriate field and laboratory studies from Unit 5. Sections 5.27 to 5.32 deal with the ecology of soil arthropods.

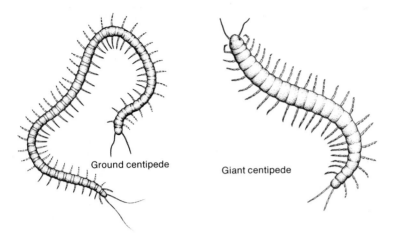

Ground centipede

Giant centipede

Fig. 3-13
Two representative centipedes. Note the pair of legs per segment. The ground centipede can be up to 10 centimeters long and the giant centipede up to 15 centimeters long.

Many-Legged Animals: Millipedes and Centipedes. Long ago, before the nature of centipedes and millipedes was well understood, these organisms were often referred to as tiny serpents. They are now classed among the arthropods and called Chilopoda and Diplopoda, respectively. The former have one pair of legs per body segment and the latter, two. No doubt their ancestors were among the first organisms to invade the land, but they have evolved only slightly over hundreds of thousands of years. Today they are still confined to humid habitats. They are included among the soil inhabitants because they are most abundant in the damp, dark soil litter of woodlands and forests.

Many legs and a reliance on moist conditions are features that centipedes and millipedes have in common. Little else shows close resemblance. Unlike millipedes, centipedes (Fig. 3-13) are predators. In their own little world they are quite efficient carnivores, to the extent that they are their own worst enemies. They are cannibals. They are efficient because they possess a set of large poison claws located just below the mouth. Glands within the claws store a poison that is lethal to small organisms which meet up with this night prowler in some dark, secluded crack or crevice. Seldom does a centipede run into enemies on its nighttime forays. Even if it does, it possesses a unique defense mechanism, sometimes used when its claws fail. When attacked and held by a leg or two, the centipede simply releases the legs in question. What is unique is the fact that the legs continue to vibrate even while unattached to the body. This feature keeps the predator's attention while the centipede beats a hasty retreat.

Millipedes (Fig. 3-14) prefer humus and rotting plant substances to flesh. Their rather weak mouthparts show none of the offensive weaponry seen in centipedes. Although their taste for plant material sometimes proves detrimental to man when

Fig. 3-14
Some representative millipedes. Note the two pair of legs per segment. Do *centi-* and *milli-* have their usual meanings when used in *centi*pede and *milli*pede?

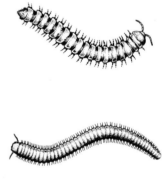

crops are involved, millipedes usually leave living material alone. They are sometimes forced to attack crops for the sake of water in very dry spells. Viewing the total picture, however, they really do more good than harm. Their bodies act like little grinding machines that fragment a great deal of litter material. This litter is often of little nutritional value, and only after it is reduced to finely powdered waste can it be used by other soil inhabitants.

When danger appears, millipedes take up a classic protective spiral pose, legs to the inside, hard shell to the outside. This pose can be used to preserve body moisture during prolonged drought, but it is most dramatic when used against common enemies like toads, birds, spiders, ants, and, possibly, some reptiles. The greatest thing going for this otherwise defenseless beast, however, is its terrible taste. A special set of organs, appropriately called *glandulae odoriferae* ("stink glands") are its ultimate weapon of defense.

Wood Lice and the Armadillo Trick. It is a known fact in nature that when one type of animal comes up with some valuable modification that aids its chances of survival in a specific environment, other animals sharing the habitat often develop the same characteristics. In biology this is called "convergence." Organisms, strictly unrelated, converge to a similar shape or behavior which is in some way beneficial to them.

Wood lice, mites, millipedes, and some mammals have all developed what may be called the "armadillo trick." Armadillos are mammals living in the southwestern United States, and Central and South America. They are active diggers and burrow-dwellers. When threatened, some have the ability to roll up into near perfect spheres with their protective shells completely enclosing them. Wood lice and some mites and millipedes can similarly protect themselves (Fig. 3-15). For wood lice this behavior can be employed not only against enemy attackers, but also on those occasions when dry surroundings threaten them. Less moisture is lost while in the coiled condition.

Wood lice are more respectfully called pill bugs, presumably because when rolled up, they are about the size and shape of a pill. They are also commonly called sow bugs. A wood louse is a crustacean and, as such, is more closely related to lobsters and shrimps than to other land arthropods such as millipedes and insects. Not surprising, then, is the fact that the pill bug fares poorly in dry air after a short time. It lacks the waxy, waterproof "skin"—the cuticle—found on millipedes and most insects. Thus it has very little control over water loss. It quickly dries out and dies in exposed dry habitats. Nevertheless,

Fig. 3-15
A wood louse can roll up into a tight ball to protect itself from predators and dry conditions.

wood lice have adapted to many areas, avoiding dryness while hiding below stones, bark, leaves, and fallen logs, or while safely tucked away in burrows that they themselves construct. In the cool of the night, the wood louse leaves its hiding place to forage on the dew-soaked leaves and other wood residues in the litter. In this way it contributes to modifying the soil.

Pseudoscorpions and Mites. Beware, a pseudoscorpion!

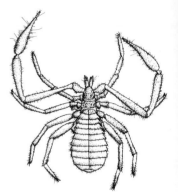

Fig. 3-16
A typical pseudoscorpion.

No call for alarm really. The only times that a pseudoscorpion lives up to its rather impressive name are in sketches drawn larger than life, and under a microscope. Even when seen magnified, it is quite clear that these little creatures, only about 1 to 7 mm long, do not have the same appearance as true scorpions. They lack the poisonous tail stinger of the real scorpion. No, it is not a scorpion. The prefix "pseudo," meaning "false," tells us that. These false scorpions are much too tiny to be a menace to anything other than tiny animals living in the soil humus.

Figure 3-16 shows why pseudoscorpions, at first glance, might be confused with their deadly distant cousins. As is the case with scorpions and spiders, pseudoscorpions have eight walking legs. They also have the claw-like arms of scorpions. These are very important in the pseudoscorpion's role as a mini-carnivore in the humus. Concealed in the darkened shadows of a rock crevice or a tiny notch in the bark of a decaying tree, the pseudoscorpion is usually ready to snap its claws on some unsuspecting passerby. A springtail would be a likely dinner dish and would be aware of its unfortunate fate only after it was too late to plan an alternate route. In a split second the pseudoscorpion catches its prey and then, through small punctures made by its claws, pumps a poison into the body of the prey from glands located in the claws. Small or immature flies, beetle larvae, ants, centipedes, mites, and even worms could just as easily perish in a pseudoscorpion's deadly grip. Luckily a pseudoscorpion is most often successful in its attack. If a young centipede or beetle larva were to get away, grow to full adult size, and then return, a pseudoscorpion would be in deep trouble. As it turns out, larger centipedes, spiders, mites, ground beetles, ants, and some birds represent the major threat to a pseudoscorpion's existence.

Pseudoscorpions do sometimes tangle with organisms many times larger than themselves. For instance, houseflies, on rare occasions, have been seen with a pseudoscorpion attached to one of their legs. The harvestman (or daddy-long-legs, as it is sometimes called) may also pick up one as a hitchhiker. Since the pseudoscorpion has rather poor eyesight, it may just be that it attacks larger organisms by mistake. On the other

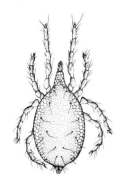

Fig. 3-17
A typical mite.

Fig. 3-18
A surface-dwelling spring-tail.

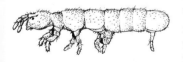

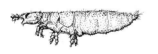

Fig. 3-19
Sub-surface springtails. From the top down, these are found at progressively deeper levels in the soil. What evidence do you see of this fact?

hand, this is an ideal way to disperse the species over large distances and into new habitats. Accident or not, a pseudoscorpion could never walk as far in its whole lifetime as it could fly air-express on board a housefly's leg in just a few minutes.

This business of hitching free rides can be taken one step farther. Pseudoscorpions sometimes carry their own "stowaways"! Little characters called mites (Fig. 3-17) are occasionally found attached as parasites to the sides of pseudoscorpions and other animals. Needless to say, they are among the smallest of the soil arthropods. Soil mites are often the most numerous of the soil arthropods, being occasionally outnumbered by springtails. They do more than just parasitize other soil organisms. Some species feed on fungi, while others dine on decomposing litter or carrion. Still others are predators, attacking tiny insects, their eggs, and their larvae. The full importance of mites is not yet known since they are so difficult to study.

Unfortunately pseudoscorpions and mites are rarely observed, even by zoologists, despite the fact they live in most areas where humus and leaf litter exist. Midgets in a miniature world, they are important and should be studied, if not for their ecological role, at least for their interesting habits and ways of life.

Tiny Wingless Insects. Picture, if you can, ants the size of elephants. A drastic picture for a human being to imagine, but one group of insects views ants just that way—as giants! The organisms are members of an extremely primitive group called the apterous (wingless) insects. Some members are less than 1 mm long when fully grown. Nearly all of these organisms spend their complete life within the soil and thus are rather important in soil ecology.

The apterous insects are broken down into a number of groups, the most important of which is the springtails. Rarely more than 4 mm long, these organisms get their name from a rather unique aspect of their anatomy. Surface-dwelling springtails (Fig. 3-18) have long appendages, including a tail which is normally held under the body. The tail filament is locked in place by a fastening device similar to a safety pin. As in a safety pin, if the locking device is released, the tail flips out from beneath, sending the animal through a tremendous leap. Remember, several centimeters to a springtail are like a 30 meter broadjump to a human!

Springtails that live immediately below the surface have tails smaller in size since jumping is normally restricted. Figure 3-19 shows a number of springtails which live at progressively deeper levels. The deepest dwellers are usually white and blind, possessing very short appendages, minus any spring-like tail. They are less than a millimeter in length.

What types of soil are most suitable for springtails? As for all the apterous insects, soil moisture is the most important factor controlling their presence. Other factors of importance include a large pore space volume in the soil. Since springtails are unable to burrow, they must squeeze through cracks and spaces already present in the earth while in search of food. This last aspect is the third major factor—food. Their diet is variable and includes parts of fungi and decomposing plant and animal material, including feces and the cast skins of other arthropods. Putting all these factors together, the top 5 to 10 cm of deciduous forest soil seem to be best for these tiny insects. Densities of 40,000 per square meter have been recorded in forest soils.

The full importance of springtails to soil formation is not really known. Certainly they have a significant role to play, being the last in a series of organisms to chew up and digest some of the organic matter. Their large numbers make up for their small size. They are one of the few types of organisms able to survive in areas where fungi are abundant. As is true for nearly every organism, their potential use as food by other organisms is important. Among their greatest enemies are predatory mites, small beetles, spiders, and fly maggots.

The remaining types of apterous insects all have in common a preference for moist habitats. Figure 3-20 shows representative members of the other groups. Little is known of the proturans. The thysanurans are sometimes called silverfish or bristletails. The diplurans have only two terminal filaments. Of the three groups, only the last has adopted predaceous feeding habits. As you can see, some of them have tail filaments which are modified as claws to capture their prey.

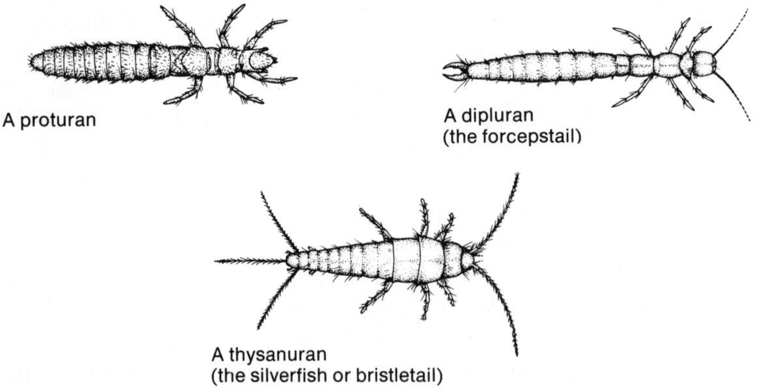

A proturan

A dipluran
(the forcepstail)

A thysanuran
(the silverfish or bristletail)

Fig. 3-20
Other apterous insects.

Ants and Termites. Have you ever stopped to consider why ants live in the ground? Not many people have. It is common knowledge that almost anywhere you stick a shovel, you will likely find ants (Fig. 3-21). But why?

If you give some thought to it you may come up with an answer, but only if you exclude food as a reason. Ants obtain nearly all of their food above ground level. The answer is protection. In the soil they are protected from predators like birds and predaceous insects that lurk above. They are also protected from the extremes of a terrestrial environment—the burning rays of the sun, the drying air, the pounding rain, and the lashing wind.

Are ants really so vulnerable that they had to become underground apartment builders? A number of facts support the idea that they are! First, the social structure of an ant colony usually requires that there be but one queen. Thus there exists but one source for reproduction. When the queen dies, so does the colony. What better way to protect the queen, the vital heart of a living colony, than to lock her away, out of sight of the dangerous and unpredictable world above? The nest becomes a fortress, easily defended, difficult to invade, and the queen is kept safe far below the surface. Second, each and every ant must pass through three totally helpless stages as an egg, larva, and pupa. At any of these stages, drastic changes in temperature or humidity would invariably spell doom. Within the ground, conditions vary from place to place, but at any one place conditions change rather slowly. Nests are constructed so that chambers exist at many different levels. If the chambers 60 centimeters deep become too cool, the larvae and pupae are dragged upstairs where the soil has been warmed by the sun's rays. If these upper chambers become too dry, the helpless young are carted off to another area where the humidity is more suitable.

It is noteworthy that nest construction varies in different regions according to differences in climate. In the northern coniferous forests, where the climate tends to result in cool springs and autumns, it is necessary for ants to capture as much of the sun's energy as possible to keep the nest warm. Only under warm conditions will the young develop properly. The wood ant (*Formica rufa*) is an example of a northern dweller. It, incidentally, is so efficient in its destruction of caterpillars, butterflies, flies, and small insects that it is protected by law in parts of Europe. This ant adds a mound of small twigs and evergreen needles as an extension to the underground section of

Fig. 3-21
One of the many types of American ants.

Fig. 3-22
Structure of an ant colony in a northern coniferous forest.

Fig. 3-23
Structure of an ant colony in a desert.

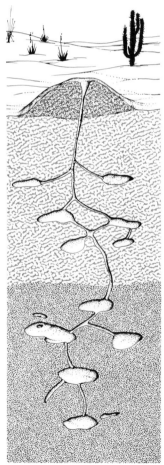

the nest (Fig. 3-22). The mound rises above the surface and thus captures much more of the slanting rays of sunlight in the morning and evening. As a result, the mound warms up quickly in the morning and the larvae and pupae are immediately transferred to the warm chambers in the mound. By noon the mound may become too warm and it may be necessary to move the young below. In the evening they may again be returned to the morning location.

In desert areas, where too much heat and not enough water are problems, ants have developed a different type of nest (Fig. 3-23). Desert species build nests that are extremely deep, descending vertically to the ground water far below the surface. At these depths, the cooler temperatures and more humid conditions are better suited for the young. In certain desert areas where valuable deposits of copper, manganese, silver, or gold occur, miners can make use of the excavated material in the ant craters to locate the possible presence of a vein.

Of what importance are ants to the soil? Despite the tremendous numbers of ants, their importance as far as physically modifying the soil is not nearly as great as you might think. Of course, the large number of tunnels and chambers must have a similar effect on the circulation of air in the soil as does the tunneling done by worms. But unlike the worms, ants keep an all-too-tidy house. The amount of humus formation that results from the presence of ants is not great. Unused organic material, including corpses of dead ants, is brought to the surface and deposited in waste piles. As a result, it is not mixed with the

Fig. 3-24
Some forms of a common termite. How is each form equipped to perform its function in the colony?

Young queen

Soldier

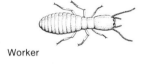

Worker

soil below. On the positive side, though, the soil particles brought to the surface may contain minerals lacking in the soil humus. Such material, easily blown over the surface, may be most useful for plant growth. This turnover, or general mixing of material, is thus of value to soil processes as is the general loosening of the soil.

A great deal more could be said concerning ants, but other books are readily available on these interesting creatures. For those who wish to investigate ants further, a research topic is included in Unit 6. It organizes a more intensive study to be done on your own.

In many parts of North America, especially where winters are not too harsh, another type of soil-dwelling social insect can be found, the termite (Fig. 3-24). These insects are commonly called "white ants," but are neither ants nor always white in color. Termites are ant-like in that their colonies consist of thousands of industrious members, each part of a single family. But their life-style is altogether different. Unlike most ants, termites are soft-bodied, and therefore avoid the rigors of life on the surface where they would be exposed to the sun's burning light and a lack of humid air. They are adapted to a life of darkness and a single product of nature, wood.

Termites thrive on cellulose, the major component in wood, and a substance which most organisms are unable to digest. Actually, if it were not for the help of millions of tiny protozoa in their digestive tracts, even termites could not digest cellulose. The use of wood as food would be impossible without these microscopic helpers. Wood also provides shelter. Whether in a fallen log, or a dead tree still standing, termites hollow out passages that never break through to the surface. If a painted board is on the menu, practically everything but the paint is consumed. This habit maintains a dark and humid place in which to live. Any form of wood might be considered a kitchen in which the hungry termite can stuff itself merely by chewing on the walls. This behavior, needless to say, has created a great dislike among humans for termites. What appears to be a solid piece of wood on the outside may, in fact, be a hollowed skeleton or shell of the real thing. As uninvited guests, these nibblers can destroy the basic framework of a house, causing irreparable damage before being discovered. Detection may come about unexpectedly some day when an arm is thrust through a weakened door, or a desk crumbles into pieces on the floor. At times like these, the average homeowner has little or no concern for the importance of termites in their natural environment. Who could blame him?

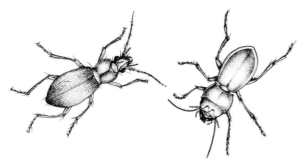

Fig. 3-25
The carabids, or ground beetles, are carnivores. How can you tell?

Fig. 3-26
The tiger beetle. What structural features make it possible for this beetle to move faster than a ground beetle?

Fig. 3-27
A scarab or dung beetle.

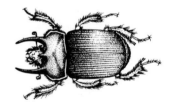

Fig. 3-28
A pair of dung beetles at work.

Yet the work of termites under natural forest conditions is beneficial. The materials locked up in fallen logs can more quickly be returned to living protoplasm after termites do their carpentry. They speed up the wood decay processes by munching wood into pieces smaller than sawdust. Other organisms, like fungi, can then more easily continue the breakdown.

Termite societies are also very interesting topics for study. For instance, the queen of a termite colony becomes a huge egg-laying machine, hundreds of times larger than any of the other termites. In some species, eggs continually drop from the queen's abdomen as workers feed her at the other end. In a single day, 30 thousand eggs might be produced. Ten million are possible in a single year. Many other facts are just as amazing. If your interest is great enough, termites are a good topic to research. Ample information is available.

Beetles. *Coleopterous*, in ancient Greek, means "sheath-winged." Scientists use the name Coleoptera to describe the insect order containing the greatest number of species of all. These insects are commonly called beetles. Over 150,000 species of beetles have been discovered.

Sheath-winged is a good description for a beetle. You might even say they are shield-winged. No other group of insects has a stronger, more armor-like skeleton than have adult beetles. As their name implies, the shielding is carried to the point where one set of wings has been modified into hard protective plates. These are held tightly over the otherwise vulnerable second set of soft wings.

The adult beetle, often looking like a miniature armored tank as it gropes about the soil surface, is better adapted to life on the soil than in it. This is not to say that adult beetles have no importance to the soil community. They do. Adult ground beetles (Fig. 3-25), called carabids, attack tiny springtails, earthworms, snails, and various other inhabitants of the leaf litter. Thus they help to maintain a balance in the numbers of other soil organisms. The animals they attack are often soft-bodied and no match for such an armor-plated adversary.

Carabid beetles remain perfectly still, waiting for potential prey to move into attacking range, but other types of beetles are more mobile. Tiger beetles (Fig. 3-26), as their name implies, can move with speed to pounce on their victims. With apparent ease, a tiger beetle's powerful jaws can rip through the skin of almost any soil inhabitant to squeeze out the edible juices inside.

Predaceous adult beetles like the tiger and ground beetles play an important role in the soil, but so do many other forms that could be termed scavengers. The scarabs, or dung beetles (Fig. 3-27), choose as a way of life to collect, store, and feed upon the droppings of various animals. First, a portion of the dung is formed into a round ball, often many times larger than the beetle itself. Then, the sphere is slowly rolled, with great strength and patience, to a proper burial site (Fig. 3-28). The beetle then begins to dig a pit in which to roll the one or two months' food supply. Often, while rolling such a great burden, the original beetle is "helped" by a second. While the first is busily preparing the pit, on some occasions the "helper" becomes so overwhelmed by the spherical delight that he carries it off for his own use, leaving the original dung beetle with an empty pit and an empty stomach.

Other scavengers include the carrion beetles or silphids (Fig. 3-29). Some of these undermine the corpses of small vertebrates, like birds or mice, and then bury them for their private food supply (Fig. 3-30). The species of scarabs and silphids that bury organic material carry out two important services. First, various flies, which might otherwise carry bacteria and diseases harmful to man, are robbed of a place for their maggots to grow. Second, organic material is quickly restored to humus once it is pulled beneath the surface.

Fig. 3-29
Some common silphids, or carrion beetles.

Fig. 3-30
Some carrion beetles undermine and then bury their prey.

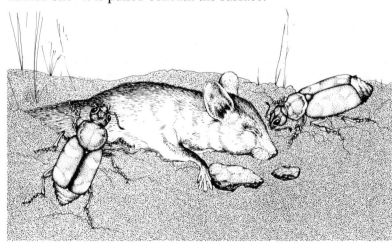

Beetle larvae are probably more important members of the soil community than the adults. Unlike the adults, the larvae are seldom protected by armor plate. Their skin is more pliable, often white or grey in color, and not nearly as resistant to water loss as the adult's. But water loss is not nearly the problem in the soil that it might be on the surface. Living within the soil, as many beetle larvae do, a more worm-like shape is of advantage for movement.

It is interesting to note that many of the species are just as predatory in the larval form as they later are as adults. Tiger beetle larvae live in specially constructed vertical burrows which open to the surface. The head and jaws of the larva neatly fill the entrance to the burrow so that no hole is visible when the larva is in its attack position. The visitor that puts its foot in the wrong place is met with sharp, powerful jaws that seize its limbs and haul it down into the darkened chamber. Special hooks on the body of the larva (Fig. 3-31) prevent it from being pulled out of its burrow. Any unfortunate guest is on a one-way, no exit, short trip.

Carabid larvae are carnivores just like the adults. However, the larvae of some beetles do not occupy the same niches as the adults. For example, the "white-grub" that is often unearthed while gardening is a scarab larva. Yet it is not a scavenger like the adult. Instead, it is a herbivore that feeds on the roots of grasses, and, all too frequently, garden plants.

It is interesting to note how unusual relationships sometimes develop between animals living in the same environment. Such a case exists between one species of rove beetles and certain types of ants. Most rove beetles are surface and litter carnivores, but this species lives underground in ant colonies. They possess hairs which secrete a substance that the ants find tasty. In return, the beetles obtain food directly from the ants' mouths, even though they are quite capable of feeding themselves. The adults take further advantage of the situation and use the ant larvae themselves as food. This injustice goes unnoticed by the ants which treat the beetle larvae exactly as their own, feeding and caring for them at all times. If it were not that both larval and adult rove beetles look and behave much like normal ants, their free-loading way of life would probably not be tolerated.

And More Insects. There remains an assortment of other insects which should be mentioned. You will likely come across some of them in your field studies.

Diptera means "two-winged" and refers to the insects commonly called flies. The larvae of flies, called maggots, are

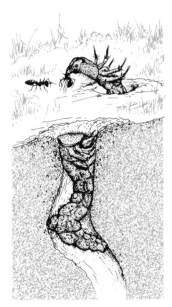

Fig. 3-31
The larva of the tiger beetle springs on its prey from a concealed burrow.

Fig. 3-32
Some soil-dwelling diptera (fly) larvae.

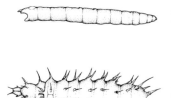

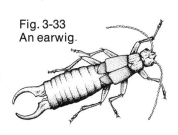
Fig. 3-33
An earwig.

Fig. 3-34
Nymph and adult stages of
the cicada.

moisture-loving creatures, unable to cope with long exposure to dry air. As a result, the larvae of some species can be found living in the water. Others are found in the rotting flesh of carrion or even the flesh of live animals. Still others live in moist plant tissues. A substantial number of fly larvae also take refuge in the soil (Fig. 3-32). Here their moisture requirements are met and food is available in the form of leaves, decaying wood, roots, fungi, carrion, and even other soil inhabitants. Their importance in moist soils may rival that of any other soil insects.

Earwigs (Fig. 3-33) received their name many centuries ago. At one time certain Europeans believed these insects would attack a sleeping person by crawling into his or her ear, and thus into the brain. Earwigs do, in fact, become most active at night. They spend much of their time exploring cracks and crevices while scavenging for food. As a result, they often enter houses through tiny cracks. More often than not they are then unable to find their way back out. No doubt, on many occasions in the past, they have horrified some sleeping giant by entering his or her ear by mistake, thus giving rise to the various superstitions associated with this insect. The European earwig has made the Atlantic crossing and is currently spreading through many parts of North America, becoming a real pest in some areas. The reasons for its dramatic success in colonization are not completely known. Certainly the maternal instinct of the female earwig has something to do with its success. After laying her eggs in a cavity in the soil, the female guards them with her life. She possesses claw-like pincers on her abdomen which give her a rather fierce appearance and no doubt make her a respectable fighter. Her young are thus given a good chance of hatching and fending for themselves.

Another insect, which you may have heard but never seen, spends up to 17 years as a soil dweller. It then makes a short and noisy appearance above ground to mate and lay its eggs. The insect is called a cicada (Fig. 3-34). It is the source of that noisy buzzing that you may have heard on a hot summer's day. The sound is something like an old electric shaver badly in need of oiling. Actually, the noise they make is necessary to attract a mate. After such a long time in the relative silence of the underground world, it is necessary to insure mating success in the relatively short-lived adult stage. During the many years below the surface, life was a leisurely affair. The nymph merely attached itself to an underground root and fed from the juicy tissues within. As an adult, the time for loafing is over, and the instinct to reproduce sparks an active finale to an otherwise dull existence.

Last of all, it should be mentioned that crickets have some minor influence on the soil. Their burrowing behavior and habit of storing green plants for food within their burrows help to mix and enrich the ground. The mole-cricket is especially well adapted to tunnel construction as can be seen in Figure 3-35. You might compare its shape to that of the mole in Figure 3-36. Their similar ways of life have given rise to remarkable similarities in appearance.

3.5   VERTEBRATES

It isn't really surprising that vertebrates (animals with back-bones) have been largely unsuccessful in exploiting the soil. It obviously takes a great deal more energy to move through the earth than to move through air or water. Any mammal is large in comparison with the cracks and crevices already present in the soil, so movement becomes a very energy-consuming task. In most instances there just isn't enough food available to replenish the energy used in burrowing. Most of the vertebrates that are classed as soil dwellers use it only for protection and shelter. Only a sparse few can scratch out a living while staying underground. Enough said about the problems. Some vertebrates have succeeded, but how?

The mole (Fig. 3-36) is an excellent example of the ceaseless effort and modifications of body structure required for the permanent resident. A mole remains active day and night, burning up energy so fast that it would starve to death in 12 hours without food. It has no choice; it cannot take the night off. With respect to modifications, the following features are a mole's ticket to success. To start with, its nose tapers to a point, giving the body a streamlined shape and offering the least resistance while burrowing. A special fur, much like velvet, covers its body. This fur will lie flat in any direction, whether the animal moves forward or backward in its burrow. Its legs are short but powerful, and its paws are tough-skinned, shovel-shaped, and supplied with claws that easily rip through the earth. Like a swimmer churning through water, the mole churns through the soil. Each stroke propels its body into a new potential food area. Each centimeter forward could reveal the odor of an earthworm or soil insect. Each centimeter could mean food. Of course, the mole will never set any land-speed records. Producing 100 meters of tunnels a day works out to 0.0039 kilometers per hour. Even so, 100 meters of loosened soil per day is a tremendous help to many of the other soil creatures so dependent on the crevices and air spaces thus provided.

Fig. 3-35
What adaptations does the mole-cricket have that make it an excellent tunneler?

Fig. 3-36
The common mole.

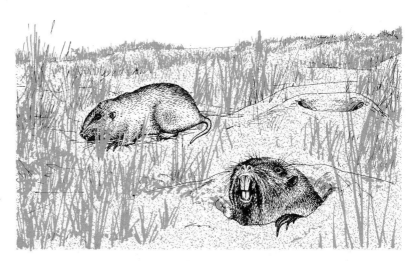

Fig. 3-37
The common pocket-gopher.

North America is the home of another group of nearly permanent underground mammals, the pocket-gophers (Fig. 3-37). Why are they called pocket-gophers? These animals possess two deep, external cheek pouches which are used to carry large amounts of food. The pouches function as huge pockets, fur-lined no less, and easily turned inside out for cleaning. As is the case with moles, the forelegs and shoulders are quite powerful, and each paw is equipped with five strong digging claws. Other adaptations include eyes and ears which are small and lips that can be closed *behind* the front teeth. This latter characteristic enables the gopher to loosen soil and rocks and cut through roots, without opening its lips and getting a mouthful of soil. Roots and tubers cut beneath the surface, and stems pulled down from above, are stored in the pockets and taken to underground storage chambers. The gopher can get all of its work done and never open its mouth, a rather admirable quality.

Other mammals that share the underground burrowing abilities of moles and pocket-gophers include woodchucks, ground squirrels, prairie dogs, chipmunks, and various species of mice. These mammals are not true soil inhabitants since much of their life is spent scampering about the surface after food. The badger is one mammal that exploits the populations of other underground species as a predator. Pound for pound it is probably the most powerful of all dirt diggers, disappearing out of sight in a matter of minutes after its smaller prey.

In North America, there are no other vertebrate types which are full-time soil residents. In other areas of the world, a few snakes and amphibians have taken to the soil, but they are of little importance.

For Thought and Research

1 (a) Describe the soil conditions which are most ideal for earthworms. Give reasons why each factor is beneficial to a large earthworm population.

(b) As you now realize, the soil is an extremely rich environment for animal and plant life. No organism lives alone, but ultimately has some direct or indirect effect on its neighbors. The activities of earthworms do a lot to modify the soil environment, making it more suitable both for themselves and other life. Describe the ways earthworms are beneficial to life in the soil. Give examples of other organisms that receive benefit.

2 (a) What soil conditions are generally preferred by nematodes? Why?

(b) You read about several types of nematodes in Section 3.2. Each of these occupies a particular niche in the soil community. How has each type adapted to fill the niche?

(c) How does the method of extraction of soil nematodes outlined in Section 5.34 make use of the fact that nematodes have certain environmental preferences?

3 (a) What soil types would you expect to have small snail populations? Why?

(b) Why do slugs and snails generally select the litter layer as a habitat?

4 Name the niches occupied by millipedes and centipedes in the soil community and give reasons for your choices.

5 Compare the structure and behavior of the terrestrial sow bug (wood louse or pill bug) with that of an aquatic sow bug. (Look up the genus *Asellus* of the order Isopoda.) Account for any differences.

6 Select any one of the insects referred to in Section 3.4. Research its life cycle. What portion of the life cycle is spent in the soil? What niches are occupied by the various forms of the insect? In the overall scheme of things, how important is it to the soil community?

7 If you have not already done so, perform the field and laboratory studies in Sections 5.22-5.34.

Recommended Readings

1 *The Forest* by Peter Farb and the Editors of *Life*, Life Nature Library, Time, Inc., 1961. Read Chapter 7, "The Hidden World of the Soil."

2 *Ecology of Soil Animals* by J. A. Wallwork, McGraw-Hill, 1970. This is probably the most up-to-date, interesting and factual book on soil life.

3 *Soil Animals* by F. Schaller, University of Michigan Press, 1968. An excellent book on the subject.

4 *Soil Biology* by A. Burges and F. Raw, Academic Press, 1967. A rather technical book dealing individually with all major soil-dwelling organisms.

5 "The Living Sand" by W. H. Amos, *National Geographic*, June 1965. An interesting account of sand-dwelling organisms.

6 *The Insects* by Peter Farb and the Editors of *Life*, Life Nature Library, Time, Inc., 1962.

7 *The World of the Soil* by Sir E. John Russell, Fontana, 1957. Contains detailed descriptions of many soil organisms.

8 *Soil Biology* by Wilhelm Kübnelt, Faber and Faber, 1961. Contains general accounts of life-styles and roles of many soil organisms.

Life in the Soil: Microorganisms

4

The soil organisms discussed in Unit 3 were all large enough to see with the naked eye. Yet, in spite of their conspicuous nature, they are neither the most important nor the most numerous of the soil dwellers. A host of *microorganisms*, organisms too small to be seen with the unaided human eye, abound in the soil. Many of these occupy the important niche of decomposer. Let us turn our attention from the obvious members of the soil community to these small but important members.

4.1 PROTOZOA

Protozoa are single-celled organisms that, a decade or two ago, were considered to be the most primitive form of animal life. Today most biologists do not class them as animals but, instead, place them with other simple organisms in a kingdom called Protista. At one time researchers thought that protozoa lived only in bodies of water, but, of the 17,000 known species, over 250 have been identified in soils. Of these 250 species, however, only a few are unique to the soil.

    Like their aquatic relatives, soil protozoa are dependent on water—not large bodies of water such as ponds and lakes, but the thin film of water surrounding soil particles and the tiny channels of capillary water trapped in the pore spaces of the soil. Populations as large as 300,000 have been counted in a single gram of soil, but usually the numbers range from 10,000 to 100,000 per gram.

Protozoa tend to be particularly abundant in the upper several centimeters of the soil and scarce in the underlying layers. Environmental factors such as aeration, acidity, and temperature undoubtedly exert some influence on the number of protozoa in a given sample of soil. However, the effects of these factors are so complex that scientists have not yet reached specific conclusions concerning them. Food supply is, however, a major factor in determining the distribution of protozoa in the soil. They dwell largely in the upper layers because most species feed on bacteria. As you know, bacteria are generally abundant in the upper soil layers where they feed on organic matter.

Although protozoa prey on a wide range of bacterial species, each protozoan species has a very clear preference. A particular species of protozoa will eat all of one type of bacterium while ignoring or even avoiding neighboring bacteria of a different type. The reasons for this are not clear, for as far as researchers can determine, most inedible forms do not secrete toxic substances that would make them less desirable to hungry protozoa. One exception is the bacterium *Serratia marcescens* which does secrete a toxin that protects itself as well as any neighboring bacteria.

A few species of soil protozoa are thought to be saprophytic, but little is known about them. Other species feed on a wide range of microflora (algae and fungi). Some species, like *Euglena*, bear chloroplasts and can make their own food. For most species, however, bacteria are the usual diet.

Once food is ingested by a protozoan, it is encased in a bubble-like structure known as a *food vacuole*. The enzymes needed to break down the food are secreted into the vacuole and digestion occurs. Useful nutrients are absorbed into the surrounding protoplasm, and any waste products are ejected from the cell or left behind as it moves on.

Under suitable conditions, soil protozoa grow and reproduce very rapidly. It is therefore understandable that they would have large appetites. One researcher estimates that a single protozoan may devour as many as 40,000 bacteria between cell divisions. Add to this the fact that it is not uncommon for many protozoa to divide at least twice a day, and you can see just how ravenous these tiny creatures must be.

Once a protozoan has grown large enough to reproduce, it usually divides asexually. The genetic material in the nucleus splits in two and the protoplasm gathers around each new nucleus. As the original cell splits, two new daughter cells take shape. But, in some cases, a more complicated sexual form of reproduction does occur. Two similar organisms join together, exchange genetic material, and then separate. In a very short time, each of these cells divides into two new organisms and reproduction is complete.

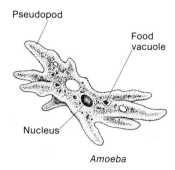

Amoeba

In spite of the variations within this group of soil organisms, their life cycles are really quite similar. They consist of an active stage during which the protozoa feed, grow, and reproduce, and of an inactive or *cyst* stage. When food becomes scarce or when adverse environmental conditions threaten the organism, the active stage is interrupted. The protozoan stops feeding, rounds itself into a sphere, and secretes a thick, resistant shell around itself. In this encysted form the protozoan can remain alive for extended periods of time. As soon as conditions improve, the protozoan emerges from the cyst and life continues as before. The transition from the encysted to the active form can take place very quickly. In a sample of dry soil there are usually no active protozoa. But if the soil is moistened, there may be as many as 10,000 active protozoa per gram within 24 hours. Researchers believe that this may be part of the reason for the remarkable fluctuations in population estimates of protozoa. Within a single day their numbers may range from hundreds to hundreds of thousands.

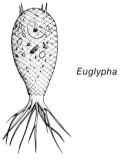

Euglypha

The protozoa of the soil can be categorized into three groups—rhizopods, flagellates, and ciliates. This grouping is based largely on appearance and methods of locomotion.

The *rhizopods* are probably the simplest of the soil protozoa (Fig. 4-1). They are microscopic or almost so, ranging in size from 10 microns to 50 microns in length (1 micron equals 0.001 mm). All rhizopods move by means of pseudopods ("false feet"). You may have already studied the amoeba, the most well-known of the rhizopods. This organism has no definite shape. Instead it changes shape constantly as its protoplasm flows this way and that in search of food. To move, an amoeba extends a portion of its protoplasm into a temporary finger-like extension or pseudopod. As the rest of the cell flows into this projection, the amoeba drifts slowly along. When an amoeba overtakes a food particle, it extends a number of pseudopods which surround the victim and enclose it in a vacuole. As you can see from Figure 4-1, some rhizopods have "shells." These species usually have much finer pseudopods than the amoeba. They stick out through a hole in one end of the shell. Other than this, the "shelled" species are very similar to the amoeba.

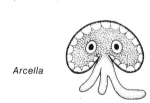

Arcella

Fig. 4-1
Some common rhizopods.

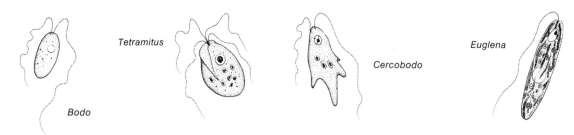

Bodo

Tetramitus

Cercobodo

Euglena

Fig. 4-2
Some common flagellates.

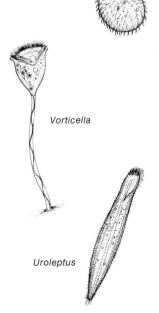

Fig. 4-3
Some common ciliates.

Colpoda

Vorticella

Uroleptus

Many soil protozoa are *flagellates*. They are much more active than the rhizopods and have a more definite shape (Fig. 4-2). Members of this group are characterized by one or more whip-like appendages known as *flagella*. These are located at one end of the organism. Although their bodies may be 3 to 10 microns long, they are often dwarfed by their flagella which may be as long as 20 microns. The lashing of the flagella helps to move the protozoan through the layer of moisture which it inhabits. A few of the flagellates, like *Euglena*, possess chlorophyll. Thus some biologists classify them as algae.

Flagellates tend to be more resistant to dry conditions than other protozoa. In fact, abundant populations of these hardy organisms have been found in the Sahara Desert.

Most soil protozoa are rhizopods or flagellates. They are frequently measured in the hundreds of thousands per gram of soil. Members of the third group, the *ciliates*, seldom number more than a thousand per gram of soil. Those found in the soil are dwarf members of common aquatic species (Fig. 4-3). The largest terrestrial ciliates are 80 microns long; their aquatic relatives may be as long as 2 millimeters. Like the flagellates, the ciliates have a distinct shape. In addition, they are covered with thousands of short hairs or *cilia* which are arranged in bands around the organism. These cilia beat in sequence and move the organism through its watery environment. Also, many of the cilia aid in feeding by creating a current of water which carries food towards the ciliate's "mouth." In contrast to the other protozoa, ciliates often have two nuclei. The larger macronucleus directs the metabolism of the ciliate, and the smaller micronucleus functions during reproduction.

Throughout the history of science, researchers have been held back by inadequate study techniques and the lack of suitable equipment. This chronic problem plagues soil scientists. Although protozoa are very abundant in the soil, study techniques are not sufficiently well developed to allow scientists to discover their functions within the soil community. Most of the knowledge concerning protozoa and their interactions with other organisms has been obtained from controlled laboratory experiments. Whether or not the results are applicable to the soil community remains to be discovered.

Since protozoa feed mainly on bacteria, it seems logical that they would have some influence on the size of bacterial populations in the soil. In instances where protozoa feed selectively on certain strains of bacteria and ignore others, those preyed upon would be diminished while those ignored would be dominant. This interaction is known to cause temporary fluctuations in bacterial populations, but it is not known if the effects are permanent. Since a close association exists between bacterial populations and soil fertility, the interaction between protozoa and bacteria could be vitally important.

If protozoa do indeed affect bacterial numbers, it is possible that they may indirectly influence bacterial transformations in the soil. In some laboratory situations, researchers have shown that an abundance of certain protozoa tends to increase the rate of nitrogen-fixation by bacteria. Scientists believe that the protozoa supply the bacteria with some type of growth-stimulating substance which increases the overall efficiency of nitrogen-fixation. But at the present time, it is not known if the same reaction occurs in the soil community.

In addition to predatory and photosynthetic protozoa, there are some that may be saprophytic. It is possible that these forms are active in decomposition processes in the soil, but their effects are unknown. Any saprophytic protozoa would have to compete with the saprophytic bacteria and fungi for what is often a limited supply of organic matter.

At the present, there is a great deal to be learned about soil protozoa. Although they exist in large numbers and are of considerable size, no definite conclusions can be drawn concerning the functions of protozoa within the soil community.

You can isolate and culture protozoa using the method outlined in Section 5.35.

4.2 FUNGI

Although bacteria are the most numerous members of the soil community, soil fungi, in their own way, are just as abundant. Each soil fungus is a great deal larger than a bacterium. In terms of the volume of organisms present, fungi and bacteria are about equal. In acid soils, fungi may outnumber bacteria. One enterprising researcher calculated that there may be as many as 100 meters of fungal filaments in a single gram of fertile soil. This is the equivalent of close to 2 miles of filaments in an ounce of soil. Other research indicates that the fungi in a single acre of soil may weigh as much as 500 pounds. However, the soil fungi are important not because of the volume they occupy, but

Fig. 4-4
Chaetomium, a representative soil fungus of the class Ascomycetes.

Fig. 4-5
Mucor, a saprophytic mold of the class Phycomycetes.

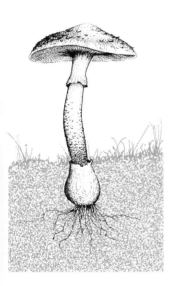

Fig. 4-6
The spore-producing structure of a common poisonous mushroom. More of this fungus is below ground than above.

because of the essential functions they perform within the soil community. In this respect they are truly equal to bacteria.

The fungi make up a very large group of organisms that are extremely varied in structure and appearance. Generally they arise from spores which germinate under suitable conditions and send out *hyphae*. These appear as white or colorless threads that may be branched or straight. The hyphae may range from 2 to 10 microns in diameter. They aggregate to form a mass known as a *mycelium*. From the mycelium new hyphae emerge and produce more spores which begin the cycle again. Usually only the reproductive or spore-producing portions of the fungus are visible. The rest of the organism is either too small to be seen or it is hidden within the soil.

Fungi have distinct physical characteristics including size, shape, and structure. Scientists have taken advantage of these traits and divided the fungi into four groups—Ascomycetes, Phycomycetes, Basidiomycetes, and Deuteromycetes (or Fungi Imperfecti).

Many species of Ascomycetes occur in the soil. Most of these produce eight spores which develop in a sac-like structure known as an *ascus*. (The Greek word *askos* means "bag.") These spores are very resistant to heat. In fact, some require a great deal of heat before they will germinate. One of the more common members of this group, *Chaetomium* is an important cellulose decomposer (Fig. 4-4). It breaks down plant residues that few other decomposers are able to handle.

The Phycomycetes are a very diverse group. You may be familiar with bread mold, which is a member of this class of fungi. The most common soil fungi in this group are saprophytic molds like *Mucor* (Fig. 4-5). They grow very rapidly and produce large numbers of spores. This allows them to take immediate advantage of any new food supply. Among the Phycomycetes are a few plant parasites that spend at least a part of their life cycle as members of the soil community. The fungus that causes potato blight is in this category.

Fungi of the third group, the Basidiomycetes, are common in the soil but they often go unnoticed. Usually they are detected only after they have produced their large spore-producing structures that appear above ground. Mushrooms are Basidiomycetes (Fig. 4-6). The part you see above ground is really the climax to a very extensive mycelial growth in the soil. Because of the difficulty in isolating the mycelia from the soil, scientists know very little about most Basidiomycetes. As with Phycomycetes, Basidiomycetes exhibit a wide range of habits. Most are saprophytes but a few are parasites.

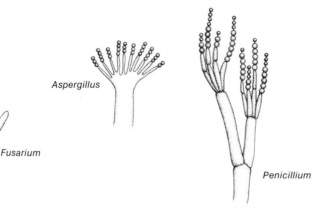

Fig. 4-7
Three Fungi Imperfecti.

Aspergillus

Fusarium

Penicillium

Many soil fungi produce large numbers of asexual spores (spores that are not the result of a transfer of materials from a male to a female organism). No sexual stage in the life cycle is known, although one may exist. Such fungi are classed as Deuteromycetes, or, more commonly, as Fungi Imperfecti (because so little is known of their life cycles). This category is quite artificial. Many researchers believe that further study of these fungi will reveal sexual stages. If this is so, this class of fungi will disappear. *Fusarium* is a common member of this class (Fig. 4-7). *Penicillium* and *Aspergillus* are probably the best-known, since antibiotics have been isolated from them.

There is much variation in the distribution of soil fungi. The distribution depends, of course, on environmental factors such as food supply, moisture content, and oxygen concentration. Generally fungi are more abundant near the surface of the soil. Here they can find adequate supplies of organic matter and oxygen. But they do exist down to depths of slightly more than one meter. Fungi at this depth must be able to survive with little oxygen. They also must be able to tolerate the higher carbon dioxide levels common in the deeper regions of the soil. Fungi generally flourish in moist conditions. But if the soil becomes too moist, the available oxygen is reduced and the fungi suffocate. Certain forms of parasitic fungi are associated only with particular host plants. The overall distribution patterns are thus quite complicated since fungi are influenced by so many environmental factors.

Some Ascomycetes and Phycomycetes are *mycorrhizal fungi*. They participate in a fascinating relationship known as *symbiosis*. They live together with the roots of a particular host plant in a relationship in which both the fungus and the plant benefit. The developing fungus surrounds the root of its host. Some of its hyphae penetrate the root while others radiate outward into the soil. The hyphae in the soil decompose organic and mineral matter and transport nutrients into the plant. At the same time, the hyphae within the root conduct sugars and other nutrients from the plant to the growing fungus. As a result of this relationship, the plant receives more nutrients than it normally would, particularly in infertile soil, and the fungus is assured of a constant supply of food. This relationship is so important to certain plants that, without it, they would perish. Some species of orchids depend on mycorrhizal fungi during the early stages of development. The fungi supply the essential nutrients until the plant becomes established. Workers in plant nurseries are careful to innoculate the soil with the appropriate fungus before they plant certain species of trees. The fungal spores remain dormant in the soil until they are activated by chemicals produced by the host plant. Then they germinate and hyphae grow toward the host and infect it.

Other types of mycorrhizal fungi do not exhibit growth on the outside of the host. Instead the hyphae grow throughout the plant and even infest the newly developing seeds. The way in which these fungi affect their host is not known, but in some cases the fungi are essential to the well-being of the plant. In fact, the seeds apparently carry the fungal infection with them to insure that the relationship will continue.

Another remarkable group of soil fungi are those that prey on other soil organisms. They are usually more abundant in soil with a high organic content and large nematode and amoeba populations. One fascinating member of this group is *Dactylella doedycoides*, commonly called the garrotte fungus. It has hyphae which develop peculiar side branches that form three-celled rings about 20 microns in diameter. When a nematode crawls through a ring, the cells swell and trap it (Fig. 4-8). There is no escape from this death-grip. Outgrowths from the ring quickly penetrate the victim and break down its internal contents. Other members of this group have equally spectacular methods of trapping their prey. They have sticky pads which hold their prey much like old-fashioned fly-paper.

Soil fungi are vitally important members of the soil community. Most of them are saprophytes. As such, they perform very essential roles in the decomposition of organic matter. Their main sources of food are the larger, more resistant organic molecules that are common in plant material. Molecules such as cellulose, lignin, and starch are resistant to bacterial decomposition but are readily decomposed by fungi. In fact, soil fungi are the primary decomposers of most woody tissues. Leaf litter, common to most forest floors, is often richly interwoven with fungal hyphae that are actively involved in this very important decomposition process.

Fig. 4-8
A garrotte fungus ensnaring a soil nematode.

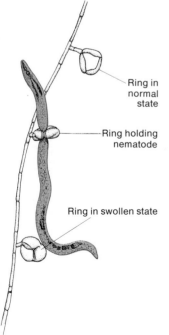

Ring in normal state

Ring holding nematode

Ring in swollen state

Fungi are also involved in the breakdown of protein substances within the soil. As a result, nitrogen compounds essential to plants are released into the soil. To some extent, fungi may be the source of other valuable nutrients. While growing, fungi incorporate many nutrients into their mycelia. When they die and decompose, these nutrients are released into the soil where they can be used once again.

In some instances, soil fungi produce chemical compounds that are quite similar to compounds found in the soil organic fraction. This has led researchers to think that soil fungi may also play an important role in the process of humus formation.

In these many ways, soil fungi assert their importance within the soil community. It is true that bacteria are more numerous and that many other soil organisms are larger, but very few can rival the importance of soil fungi in the soil ecosystem.

4.3   BACTERIA AND ACTINOMYCETES

Bacteria. Although they are invisible to the naked eye, bacteria are the most important members of the soil community. If they ceased to function, the surface of the earth would soon be dramatically altered. Debris consisting of dead plants and animals would pile up, day after day, until life as we know it ceased.

With the exception of viruses, bacteria are the smallest living things. The average soil bacterium is about 0.5 microns in diameter and from 1 to 3 microns long. Bacteria are so small that 250,000 of them could occupy the period at the end of this sentence. Their size, or perhaps we should say their lack of it, makes bacteria very difficult to study.

Under an ordinary light microscope bacteria are barely visible. They appear as single cells of varying shapes. Some are spherical, some are rod-like, and some are spiral-shaped (Fig. 4-9). Individual bacteria may assume different shapes depending on their age and their immediate surroundings. The single cells may remain isolated from one another or they may be clumped together to form colonies. Still others may join end to end to form long chains.

With the aid of powerful electron microscopes, scientists have discovered that each bacterial cell has a definite cell wall similar to the wall that surrounds a plant cell. Many bacteria have whip-like flagella which beat to and fro to help the bacteria

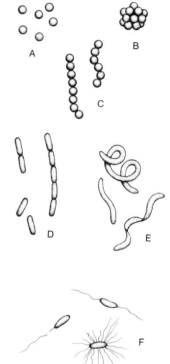

Fig. 4-9
Common types of bacterial cells. The spherical cells are known as cocci. They occur (A) singly, (B) in clusters, and (C) in chains. The rod-shaped cells (D) are called bacilli. The spiral-shaped cells (E) are called spirilla. (F) shows some flagellated bacterial cells.

move through moisture in the soil. Although they lack definite nuclei, bacteria do have accumulations of genetic material that resemble the nuclei of other cells. Some bacteria secrete a sticky coating or capsule around themselves. The true function of this is not known, but some scientists believe that it affords protection from predators. Also, if these capsule-forming bacteria are present in large enough numbers, they play an important role in determining soil structure by cementing together mineral particles and humus.

Bacteria have tremendous reproductive capabilities. Under ideal conditions a single bacterium could, in 24 hours, produce a colony large enough to cover an entire football field with a layer 30 centimeters thick. Some bacteria simply divide to create two new cells, while others undergo a complex sexual process to produce more bacteria. Many bacteria can form spores whenever living conditions become unfavorable. These spores are highly resistant to extremes of temperature and moisture. When conditions improve, the spores germinate and new bacteria emerge. As a result of these abilities, bacteria can take immediate advantage of any new food supply.

The most astonishing characteristics of soil bacteria concern their chemical nature rather than their physical appearance. They are quite varied with respect to the chemicals they require to live and their ability to perform complex chemical reactions.

Most of the soil bacteria require an abundant supply of free oxygen. Such bacteria are known as *aerobes*. On the other hand, certain of the soil bacteria, the *anaerobes*, flourish without free oxygen. Aerobes get their energy by breaking down organic matter through oxidation. A third group of bacteria, called the *facultative anaerobes*, combine the best of both worlds. They can function either aerobically or anaerobically.

In order to survive, living organisms must have adequate supplies of energy and elements like carbon. Most soil bacteria obtain both of these from one source—the organic compounds that make up dead plant and animal material. These are known as *heterotrophic* bacteria. A second group, the *autotrophic* bacteria, obtain these necessities in an entirely different manner. They get all the carbon they need from free carbon dioxide. In this respect the autotrophs resemble green plants. Their energy is derived from various sources, including inorganic compounds and light.

The heterotrophic bacteria are largely responsible for the decomposition of dead organisms in soil. This lengthy process is carried out in a series of steps with specific bacteria participating in each step. The overall result is that large organic molecules are broken down into molecules small enough to be absorbed by plants or released into the atmosphere. In this way, soil bacteria perform a major role in the recycling of essential nutrients within the biosphere.

Soil bacteria are important in the recycling of nitrogen, an element that is essential to all green plants. Certain species of bacteria, known as nitrogen-fixers, have the ability to remove free nitrogen from the atmosphere and form nitrogen compounds that can be absorbed from the soil by plants. These nitrogen-fixing bacteria are found in nodules (lumps) on the roots of legumes like peas, beans, and clover (Fig. 4-10). The relationship between the bacteria and the plant is an example of symbiosis—a mutually beneficial relationship between two organisms. The plant provides the bacteria with protection and a steady supply of food, while the bacteria supply the plant with the much-needed nitrogen compounds.

Fig. 4-10
Root of a soybean plant showing the nodules that contain nitrogen-fixing bacteria.

Recently scientists have discovered another important task that bacteria perform in the soil. Certain heterotrophic soil bacteria appear to be the only natural agents that can decompose the insecticides and other organic poisons that man releases into the environment. They can even break down some compounds that scientists can break down only with high temperatures and complex equipment.

In spite of their small size, soil bacteria perform unique functions in the soil ecosystem. They must be recognized as the most important group of organisms within the soil community.

Actinomycetes. Actinomycetes occupy a position midway between bacteria and fungi. They possess characteristics of both groups, plus a few interesting traits of their own.

Actinomycetes look much like long-branched bacteria. In fact, it is difficult to distinguish between a single bacterial cell and a fragment broken from an actinomycete. In addition, actinomycete spores are identical to those of bacteria. The resemblance between these two types of organisms is not only physical. Actinomycetes and bacteria share many similar physiological and chemical characteristics. Their reactions to chemical stains and dyes are often similar. And, in keeping with their bacterial heritage, actinomycetes also fall prey to the same types of viruses that attack bacteria, the phages (see Section 4.4).

In other ways, the actinomycetes resemble soil fungi. The single-celled actinomycetes produce extensive mycelia. Each mycelium is made up of many minute threads or hyphae which range from 0.5 to 1.2 microns in diameter (Fig. 4-11). This diameter is about the same as that of many soil bacteria.

These tiny organisms are the second most abundant members of the soil community. They are responsible for the typical earthy smell of most soils. Because of the minute size and the delicate nature of actinomycetes, scientists find it very difficult to arrive at accurate estimates of their population size. However, estimates in the neighborhood of 100,000 to 100,000,000 organisms in a single gram of soil are not uncommon. They tend to be most abundant in the upper reaches of the soil but, in some cases, as many as 100,000 actinomycetes have been found in a gram of soil from the C horizon.

Most actinomycetes are saprophytes. They feed on dead organic matter and play an essential role in the process of decomposition. But they do not become active until bacteria and fungi have completed their tasks. Once the competition from these organisms lessens, the actinomycetes continue the decomposition process by breaking down the remaining resistant compounds. Many actinomycetes have the ability to function at

Fig. 4-11
The mycelium of *Streptomyces*, one of the most common soil actinomycetes. Note the coiled spore chains.

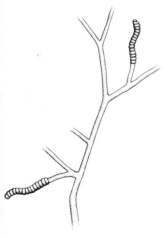

high temperatures. This gives them a definite advantage over most soil bacteria, which are incapacitated by excessive heat.

A few actinomycetes are parasitic. These usually infest plants. They cause such soil-borne diseases as potato scab and sweet potato pox.

A large number of soil actinomycetes have the ability to produce chemicals toxic to many other soil inhabitants. You might say that they wage a type of chemical warfare. This is quite common among members of the genus *Streptomyces*. Some species of this genus are grown commercially in laboratories to provide supplies of such important antibiotics as streptomycin. The importance of these antibiotics in the natural soil community still has not been established. However, it seems logical that the presence of certain actinomycetes controls the growth of populations of other microorganisms in the soil.

4.4 VIRUSES

Viruses are truly remarkable. They appear to occupy a position midway between the living and the non-living. In fact, scientists have not yet decided on the true nature of these "things." Some argue that they cannot be living organisms because they lack the typical cellular structure of other living things. Instead they look more like some form of machine (Fig. 4-12). They have no metabolism of their own and appear to show no response to external stimuli. They are so simple that they can be crystallized in much the same manner that salt can be crystallized from its solution. Outside of living cells viruses show no characteristics that we commonly associate with living organisms. Yet within the living cells of a host organism, viruses exhibit one of the basic life processes—they produce more of their own kind. The manner in which they do this is most intriguing. They take over control of the host cells and force those cells to use their own nutrients to make new viruses instead of new cellular material. Because viruses can reproduce, some scientists class them as living organisms. Regardless of which point of view they hold, all scientists recognize that viruses have important effects on more complex organisms. For this reason, they are usually included in any discussion of microorganisms.

Viruses can be observed only with an electron microscope. Magnifications of about 50,000 times are required just to see the shapes of viruses. Magnifications of up to 500,000 times are commonly used to see structural details. As a result of such photographs and numerous chemical tests, scientists believe that

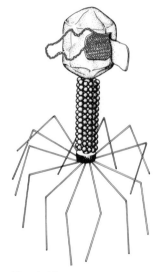

Fig. 4-12
Diagrammatic representation of a phage virus. What does it look like to you?

Fig. 4-13
General structure of a
virus that attacks plants
and animals.

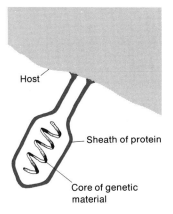

Host

Sheath of protein

Core of genetic
material

viruses consist of an outer sheath made of protein and an inner mass of genetic material which contains the blueprint for more viruses.

Soil viruses can exist in the soil in an inactive crystal stage for long periods of time. Remember, though, that they can reproduce only inside living cells. A virus injects its genetic material into the host cell; this material uses the metabolic machinery of the cell to produce numerous new viruses; the cell then bursts and the new viruses are released into the environment. It is at this stage that damage to the host occurs.

Some viruses can reproduce only in one particular type of cell, while others can multiply in a wide range of hosts. Generally, they can be divided into three groups on the basis of host preference. There are viruses that parasitize plants, others that parasitize animals, and a third group that parasitizes microorganisms. The viruses that attack animals and plants have the simplest shape. They are essentially rod-like in appearance (Fig. 4-13).

The viruses that parasitize bacteria and actinomycetes are known as *bacteriophages* or, simply, *phages* (see Figure 4-12). They have a more complex structure than the plant and animal viruses, and look somewhat like a lunar landing craft. They consist of a protein head with a diameter of 0.05 to 0.10 microns and a tail which is narrower, but may be as long as 0.2 microns. The genetic material is stored in the head of each virus. It is injected into the host through the tail of the virus much as antibiotics are injected with a syringe.

One very important group of phages are those that parasitize the *Rhizobium* bacteria. These bacteria are the nitrogen-fixing bacteria. If they become infected with a phage virus, the entire nitrogen cycle is disrupted.

The complete role of viruses in the soil community is still unknown. Scientists have studied individual cases, but even these studies tend to be incomplete. Scientists know how a single phage affects a single bacterium, but they do not know how millions of viruses affect the bacterial population of the soil.

For Thought and Research

1 (a) Why are protozoa most abundant in the upper few centimeters of the soil?

(b) Why are soil protozoa generally smaller than aquatic species?

(c) Lay out a food chain involving soil protozoa. Include at least five trophic levels in this chain.

(d) What advantage does the flagellate *Euglena* have over other soil protozoa? How deep in the soil would you expect to find this organism? Why?

2 (a) What niches can be occupied by soil fungi? Describe an example in each case.

(b) What do you think would ultimately happen to the soil of a deciduous forest if the soil fungi were all killed?

(c) Construct food chains showing soil fungi in the various niches named in (a).

(d) Consult a biology book for details on how to culture bread mold. After your culture has grown, use a microscope to identify the hyphae, mycelia, sporangia, and spores. The general structure of this mold is similar to that of many soil fungi. You could also examine leaf mold and the mold on oranges, if you wish to see examples of different types of molds.

3 (a) Why are bacteria necessary components of the soil community?

(b) Find out the meaning of the term "chemosynthesis." Compare this process to photosynthesis. In what ways are each of these important to the soil ecosystem?

4 (a) Why do farmers sometimes grow a crop of beans or clover on a piece of land one year before growing a crop of corn on the same land?

(b) How can the presence of actinomycetes in the soil be easily detected? What type of soil do they usually inhabit? Collect samples of soil from various sites and verify your answers to these questions.

(c) What niches do bacteria and actinomycetes occupy in soil ecosystems?

5 (a) Some scientists say viruses are living; others say they are non-living. Could both groups be correct? Discuss.

(b) Find out what is meant by "genetic material."

(c) What effects do you think viruses have on the soil ecosystem?

6 Populations of algae are found in most soils. Consult the *Recommended Readings* to find out such things as the niche occupied by algae, their relative abundance in various types of soil, the depth to which they can be found, and the types of algae that inhabit soil.

7 Perform the field and laboratory studies in Sections 5.35-5.39.

Recommended Readings

The following books contain further information on soil microorganisms.

1 *Micro-organisms in the Soil* by A. Burges, Hutchinson University Library, 1958.

2 *The World of Soil* by Sir E. J. Russell, Fontana, 1957.

3 *Life in the Soil* by R. M. Jackson and F. Raw, Edward Arnold (Publishers), 1966.

4 *Introduction to Soil Microbiology* by M. Alexander, John Wiley & Sons, 1961.

5 *Soil Animals* by F. Schaller, University of Michigan Press, 1968.

Field and Laboratory Studies

5

5.1   SOIL SAMPLING

In many of the exercises of this Unit, a number of soil samples are required. As these samples are removed, they should be placed in plastic bags, sealed, and labeled. A general description of the area from which samples are taken should also be written.

Several methods of removing samples are available. The method used depends on the depth of the sample required, the size of the sample, and the frequency with which a site is visited.

Method 1 THE SOIL PIT

If soil studies are carried out by several classes on a large school property, one or more permanent pits could be dug in inconspicuous spots where the soil profiles are different. These pits should be about 1½ meters wide and 1½ meters deep at the deepest point (Fig. 5-1). They can be used for observation of soil profiles and for obtaining samples when required. In this way, time is saved if sampling is done during a school period, and the school's landscaping does not suffer as much as it would if many samples were taken all over the property.

Method 2   SOIL SAMPLERS AND AUGERS

Several instruments have been designed for the removal of small samples from various depths. The most obvious ones are the ordinary shovel and the garden trowel. Several other more convenient instruments have been designed.

The soil auger (Fig. 5-2) screws down into the soil to the depth required to cover the threads. When it is pulled out, the soil in the threads is removed as a sample. The auger can then be placed in the same hole and an additional number of centimeters of soil removed.

The core sampler is the most convenient device for sampling damp soils which are not stony. The sampler is slowly pushed into the ground and the depth noted. When the sampler is pulled out, the soil core is removed through the base of the sampler. The sampler can be reinserted in the same hole to go to a greater depth.

Fig. 5-1 *(Left)*
A soil pit.

Fig. 5-2 *(Right)*
The bottom of most soil augers can be removed and replaced with a core sampler.

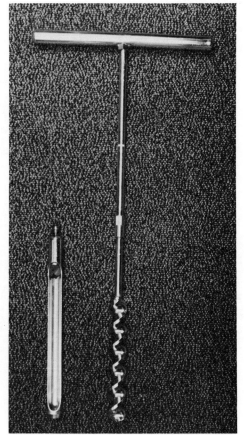

A post-hole auger (Fig. 5-3) works much like a combined soil auger and core sampler. The hole that it leaves is wider and each sample taken is usually shallower. As a result, it takes much longer to get a sample from a given depth.

The main disadvantage of these samplers is that samples can be taken only up to the depth of the handle of the device. For samples from any greater depth, a pit must be dug.

5.2 SOIL TEMPERATURE

Soil temperature varies with depth, season, soil type, time of day, light intensity, and moisture content. The temperature range in soil affects the distribution of organisms within it. As a result, temperature readings can assist you in analyzing why a population of organisms is found in a particular soil.

The following method will help you to compare temperatures taken at various sites.

Materials

soil thermometer (Fig. 5-4)

Procedure

a) Select a site where the depth of the various soil layers is known.

b) Read the air temperature and record it.

c) Read the soil temperature at the soil surface and at depths of 5 cm, 10 cm, 15 cm, and 20 cm. Record the results.

Fig. 5-3
A post-hole auger is slow, but it leaves a hole large enough to peer into.

Fig. 5-4
Two types of soil thermometer.

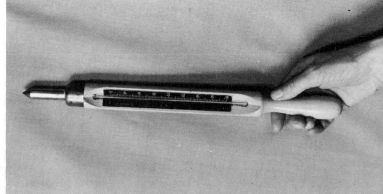

d) Repeat this procedure on several days in the morning, at noon, and in the late afternoon.

e) Repeat this procedure to compare shady and open areas, north and south slopes of a hill, areas where conditions are wet, dry, or damp, and areas where soil types differ. In doing this, be sure that all variables except the one being observed are constant. For example, when comparing the temperatures on north and south slopes, make sure that the soil types, the moisture content, the depths, the light intensity, and the time of day are approximately the same.

Discussion

Do the soil and air temperatures vary greatly? How did the soil temperature vary with depth? Was there variation in your readings with regard to the time of day they were taken? Did the amount of shade affect the temperature at all levels of soil? In which types of soils did you observe the greatest temperature fluctuation? The least fluctuation? How does moisture content appear to affect soil temperature?

5.3 MOISTURE CONTENT OF SOIL

The amount of moisture found in soil varies greatly with the type of soil, the climate, and the amount of humus in that soil. The types of organisms which can survive in an area are largely determined by the amount of water available to them, since this water acts as a means of nutrient transport and is necessary for cell survival.

 The following method for measuring the moisture content of soil involves comparing the weight of a soil sample before and after it has been dried in an oven. From this information, the percent of moisture can be calculated.

Materials

a) large beaker (or evaporating dish)

b) filter paper

c) soil (50–75 gm)

d) oven (temperature range should be such that readings near 100°C are accurate)

Procedure

a) Weigh the beaker and record its weight.

b) Add the soil to the beaker and reweigh it. Record the weight.

c) Heat this in an oven set at $100\,^{\circ}C$ for approximately 24 hours or until the weight is constant.

d) Reweigh the beaker filled with the dried soil. Record the weight.

Calculations

Using the following relationships, calculate the percent of moisture in the soil sample.

a) $$\frac{\text{weight of beaker} + \text{original soil} - \text{weight of beaker}}{\text{weight of soil sample before drying}}$$

b) $$\frac{\text{weight of beaker} + \text{dried soil} - \text{weight of beaker}}{\text{weight of dried soil}}$$

c) $$\frac{\text{weight of soil sample before drying} - \text{weight of dried soil}}{\text{weight of water in soil sample}}$$

d) $$\frac{\text{weight of water in soil sample}}{\text{weight of dried soil}} \times 100 =$$

percent of moisture in the soil sample

Discussion

Why was the weight of the dried soil rather than the weight of the original soil used in the relationship in d)?

Attempt this method on various types of soils from several locations and account for the differences in your results.

Samples from different depths at the same location should also be analyzed for moisture content. Account for your results.

5.4 WATER-HOLDING CAPACITY OF SOIL

The water-holding capacity of soil is mainly dependent upon two things—how much humus is in the soil, and the size of the soil particles. Although some soils can absorb their own weight or more in water, ideally the water content should be only 60–80% of the soil's capacity. If soils contain less than 60% of their capacity for water, there is not enough water for the cellular needs of many organisms. If soils contain more than 80% of their capacity for water, there is too little oxygen available for the growth and activity of many microorganisms.

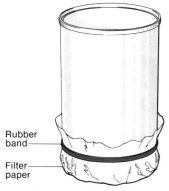

Can with both ends removed

Rubber band

Filter paper

Fig. 5-5
Determining the water-holding capacity of soil.

Materials

a) a can with both ends removed, one end covered with filter paper, held in place by a rubber band (Fig. 5-5)

b) sample of oven-dried soil

Procedure

a) Weigh the can with the attached filter paper. Record the weight.

b) Place the oven-dried soil in the can. Reweigh it.

c) Slightly moisten the filter paper on the end of the can.

d) Weigh the completed apparatus. Record its weight.

e) Set the can (filter paper end down) in water, so that the lower half is immersed. Leave it for 14 to 16 hours (overnight).

f) After this time, remove the can from the water. Transfer it to a rack where it can drain for approximately 30 minutes.

g) Wipe the surface of the can dry and weigh it. Record the weight.

Calculations

Using the following relationships, calculate the percent water-holding capacity of your soil sample.

a) $$\frac{\text{weight of water-soaked soil } + \text{ apparatus} - (\text{weight of oven-dried soil } + \text{ apparatus } + \text{ wet paper})}{\text{weight gained by water absorption in soil}}$$

b) $$\frac{\text{weight of dry apparatus } + \text{ oven-dried soil} - \text{weight of dry apparatus}}{\text{weight of oven-dried soil}}$$

c) $$\frac{\text{weight gained by water absorption in soil}}{\text{weight of oven-dried soil}} \times 100 = \text{percent water-holding capacity of the soil}$$

Discussion

How much water should have been added to your sample of oven-dried soil for it to be at 60% to 80% of its water-holding capacity? In what situations would knowing the water-holding capacity of soil be useful?

Try this method on several types of soil and account for the differences in your results.

5.5 COMPARISON OF THE WATER-HOLDING CAPACITIES

Alternatives to expressing the water-holding capacity of soil as a percent of its dry weight are available. One such method is to compare the water-holding capacity of two or more soils which differ in humus content and particle size.

Method 1

Materials

a) cans with both ends removed (all cans must be the same size)
b) filter paper or cloth
c) rubber bands
d) racks
e) large mayonnaise jars or mason jars (the mouth of the jar must be larger than the base of the can)

Procedure

a) Using a rubber band, secure a piece of filter paper or cloth to the base of each can.
b) Oven-dry each sample of soil. (Try to get a sample from a field where grasses or legumes have recently been grown, a sample from a denuded playground, a sample from a garden, and so on.)
c) Fill each can approximately ⅔ full of its soil sample.
d) Place the cans on racks over the mouths of the jars (Fig. 5-6).
e) Pour equal amounts of water into each can.
f) Measure the time which elapses before water begins dripping into each jar; how long the water continues to

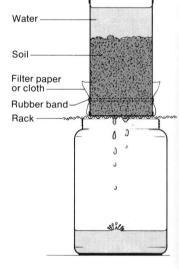

Fig. 5-6
Assembled apparatus for comparing the water-holding capacity of soils.

drip into each jar; and how much water eventually passes through each soil.

Method 2

Materials

a) glass beakers of same size (for example, 250 ml)

b) various soil samples (for example, clay, loam, sand, silt)

c) 100 ml graduated cylinder

Procedure

a) Assign 1 beaker to each soil type. Fill it up to 2 cm from the top with the soil.

b) Using the graduated cylinder, add as much water to each beaker as can be absorbed. Note the volume absorbed in each case.

Discussion

Which soil sample absorbed the most water? Why? Why would some soil samples be unable to absorb large amounts of water?

5.6 PERCOLATION RATE: SPEED OF INFILTRATION INTO SOIL

This exercise is one which can be performed in the field. It involves the same principles as used in Section 5.5.

Materials

cans of the same size with both ends removed

Procedure

a) Work each can into the ground about the same distance (about 2 cm).

b) Pour a measured amount of water into each can.

c) Measure the time required for all the water to enter the soil. Record it for each can and average the results.

d) Repeat this procedure in several types of soil—in a lawn, on a dirt road, in a garden, in a deciduous woods, and so on.

e)   If worm burrows are available, do one measurement with the mouth of a burrow in the center of the can.

Discussion

Which soil types absorbed the water most rapidly? Why? State the advantages and disadvantages of fast infiltration of water into the soil.

Attempt this exercise using muddy water. Account for the difference in your results. How does this apply to percolation rates and erosion? How does the activity of worms help prevent flooding?

5.7   CAPILLARITY OF SOIL

Percolation of water into the soil carries water with its dissolved and suspended materials into the depths of the soil. Evaporation dries out the upper layers of the soil. In areas where there is little rainfall, this should mean that few organisms can survive in the upper layers of soil. However, we know that organisms do live in the upper layers of soil in low rainfall areas. Where does the water which is necessary for their survival come from? The following exercise may provide some answers.

Materials

a)   ring stand and clamp
b)   rubber band
c)   large beaker (1 liter)
d)   ruler
e)   clear colorless tube, at least 5 cm in diameter and 15 cm long
f)   200 ml soil samples (fine sand, coarse sand, clay)
g)   filter paper

Procedure

a)   Seal off one end of the tube by fastening filter paper to it with a rubber band.
b)   Slowly add approximately 200 ml of clay to the tube.
c)   Assemble the apparatus as shown in Figure 5-7.
d)   Slowly lower the tube until its base is just below the water surface.

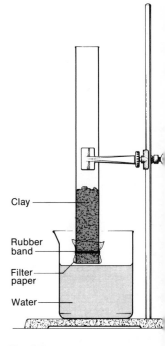

Clay

Rubber band

Filter paper

Water

Fig. 5-7
Demonstrating capillarity in soil.

e)   Using the ruler, measure the height to which water rises in the clay at 30-second intervals. Make your measurements from the base of the tube.

f)   Repeat this procedure using fine sand and then coarse sand.

Discussion

What is capillarity? Why is it important for the survival of soil organisms? If you had two gardens, one in which the soil was mostly sand and the other with mainly clay soil, which one would you have to water more often? Why?

5.8   PORE SPACE OF SOIL

In every sample of oven-dried soil, a certain percent of its volume is occupied by air. This is called the pore space and it is found between the solid soil particles. In moist soil, some pore space is filled with water and some with air. The amount of pore space varies from one type of soil to another. In the following exercises, you will measure the pore space in two soil samples and compare them.

Method 1

Materials

a)   can about 5 cm deep and 8-10 cm in diameter, with both ends removed (for example, large tuna fish can)

b)   filter paper

c)   2 rubber bands

d)   spatula or knife (some instrument suitable for leveling)

e)   soil samples (for example, coarse sand, clay)

f)   razor blade

g)   oven, accurate in the range near 100°C

Procedure

a)   Oven-dry the soil samples for 24 hours in an oven set at 100°C.

b)   Measure the internal diameter and height of the can. Record the measurements.

c)   Fix the filter paper to one end of the can using a rubber band.

d) Place the second rubber band right at the base of the can. (This insures that no soil finds its way up the sides of the can beneath the filter paper.)

e) Weigh the can and dry filter paper. Record the weight.

f) Wet the filter paper and reweigh the apparatus. Record the weight.

g) Add 8 to 10 grams of soil at a time to the can. Tap the top of the can after each addition, to settle the soil.

h) When the can is full, level off the top and then tap the can. Continue to add soil to the top and level it off until only very slight settling occurs when the can is tapped. Top off the can and level it.

i) Place the can in 1–2 cm of water. Leave it overnight.

j) By morning, a portion of the soil will have expanded above the top of the can. Remove this with a razor blade drawn across the top of the can at an angle.

k) Weigh the apparatus containing the remaining wet soil. Record the weight.

l) Dry the soil in the can for 24 hours in an oven set at 100°C.

m) Cool the can and its contents. Weigh the apparatus and record the weight.

n) Repeat the procedure with one or more other types of soil.

Calculations

For each sample calculate the percent pore space as follows:

a) $$\frac{\text{weight of can + wet paper + wet soil} - (\text{weight of can + wet paper})}{\text{weight of wet soil}}$$

b) $$\frac{\text{weight of can + dry paper + dry soil} - (\text{weight of can + dry paper})}{\text{weight of dry soil}}$$

c) $$\text{internal volume of can} = \frac{2}{3}\pi\left(\frac{\text{internal diameter of can}}{2}\right)^2 \times \text{height of can}$$

d) $$\frac{\text{weight of wet soil} - \text{weight of dry soil}}{\text{internal volume of can}} \times 100 =$$

$$\text{percent pore space}$$

Method 2

Materials

a)   2 graduated cylinders (100 ml)

b)   stirring rod

c)   2 oven-dried soil samples (coarse sand, clay)

Procedure

a)   Fill one of the graduated cylinders with coarse sand up to the 55 ml mark.

b)   Pack the soil by tapping the cylinder bottom quickly for ½ minute on your hand.

c)   Record the volume of the soil.

d)   Pour 70 ml of tap water into the other cylinder.

e)   Pour the measured coarse sand into the water.

f)   Stir it with the stirring rod. To let all the air escape, let it stand for 5 minutes.

g)   Record the volume of this mixture.

h)   Repeat the procedure for clay.

Calculations

When the soil was added to the water, water filled the pore spaces. Calculate the percent pore space using the following relationships:

a) $$\frac{\text{volume of soil } + \text{ volume of water} - (\text{volume of soil } + \text{ water mixture})}{\text{pore space volume}}$$

b) $$\text{percent pore space} = \frac{\text{pore space volume}}{\text{soil volume}} \times 100$$

Discussion

Which soil type has the largest percent pore space? Account for this.

State several reasons why air is necessary in soil.

How is water withdrawn from pore spaces under normal conditions?

How might rainstorms or changes in atmospheric pressure bring "fresh" air to the soil?

5.9 SEPARATION OF SOIL ACCORDING TO PARTICLE SIZE

Soils tend to separate on the basis of particle size. The following exercise is a simple method for demonstrating this principle.

Materials
a) large mayonnaise jar
b) sample of mixed soils from the same general area

Procedure
a) Fill the jar approximately $\frac{2}{3}$ full of water.
b) Pour in the soil until the jar is approximately half-full of soil.
c) Agitate the jar vigorously for at least 30 seconds. Then let it stand until the soil settles.

Discussion
Describe and account for the appearance of the settled soil. In particular, why do particles having the same density but different sizes settle at different rates? Describe and account for the appearance of the water above the soil. How does the principle illustrated in this exercise apply to the natural formation of soils?

5.10 DETERMINATION OF SOIL ORGANIC CONTENT BY IGNITION

The organic content of soil greatly influences the plant, animal, and microorganism populations in that soil. Decomposing organic material provides many necessary nutrients to soil inhabitants. Without fresh additions of organic matter from time to time, the soil becomes deficient in some nutrients and soil populations decrease.

In this exercise, two methods of determining soil organic content by ignition are outlined. Organic matter is made of carbon compounds which, when heated to the correct temperature, are converted into carbon dioxide and water. In the ignition process, a soil sample is heated to a high temperature. The organic matter in the soil sample is given off as gases. This results in a change in weight which allows you to calculate the organic content of the sample heated.

Method 1 is the preferred method because the temperature range can be controlled to prevent unwanted chemical changes. Method 2 is to be used only if you do not have a suitable oven.

Materials

a)   size 0 crucible

b)   oven with temperature range including 500°C–700°C

c)   tongs

d)   soil sample (approximately 10 grams)

Procedure

a)   Oven-dry the soil sample (see Section 5.3).

b)   Weigh the crucible and record the weight.

c)   Place the soil sample in the crucible and reweigh it. Record the weight.

d)   Place the crucible and soil in the oven at 500°C to 700°C. Leave it for 2 hours.

e)   Remove the crucible from the oven, using tongs. Allow the crucible to cool.

f)   Weigh the cooled crucible and its contents. Record the weight.

Materials

a)   crucible and lid

b)   clay triangle

c)   Bunsen burner

d)   ring stand

e)   soil sample (approximately 10 grams)

Procedure

a)   Oven-dry the soil sample (see Section 5.3).

b)   Weigh the crucible and its lid. Record the weight.

c)   Place the soil sample in the crucible and cover it with the lid. Reweigh them and record the weight.

d)   Using the clay triangle and ring stand, set the crucible with its soil over the Bunsen burner.

e)   Place the lid partially over the crucible so that the fumes can escape, but not the soil particles. Heat the crucible strongly until no more visible fumes are given off.

f)   Cool the crucible. Weigh the crucible, lid, and contents. Record the weight.

Calculations (for either method)

Using the following relationships, calculate the percent organic content of the soil sample.

a) $$\frac{\text{weight of crucible } + \text{ soil before heating} - \text{weight of crucible}}{\text{weight of soil before heating}}$$

b) $$\frac{\text{weight of crucible } + \text{ soil after heating} - \text{weight of crucible}}{\text{weight of soil after heating}}$$

c) $$\frac{\text{weight of soil before heating} - \text{weight of soil after heating}}{\text{weight of organic matter in soil sample}}$$

d) $$\frac{\text{weight of organic matter}}{\text{weight of soil before heating}} \times 100 =$$

                                              percent organic matter

---

Discussion

Carry out this investigation using soil samples from a variety of locations. Account for your results.

Also perform the study on soil from several different depths at the same location. Account for your results.

The total organic matter in soil can also be calculated using colorimetric analysis. This technique is outlined in *Laboratory Manual of General Ecology*, G. W. Cox (Wm. C. Brown Co., 1967).

5.11   COMPARISON OF SOIL ORGANIC CONTENT USING HYDROGEN PEROXIDE

A soil rich in organic material is usually dark in color because of the carbon content. Hydrogen peroxide oxidizes carbon compounds to produce carbon dioxide. What effect would you expect this to have on the color of the soil?

Materials

a)   4 beakers (250 ml)

b)   2 soil samples from the same site, one from the top 10 cm of soil and the other from a depth of at least 50 cm

c)   6% hydrogen peroxide

Procedure

a)   Label the beakers A, B, C, and D.

b)   In beakers A and B, place approximately equal quantities of soil from the top 10 cm of the site.

c)   In beakers C and D, place approximately equal samples taken from deep in the soil.

d)   To beakers A and C, add enough water to cover the soil.

e)   To beakers B and D, add enough 6% hydrogen peroxide to cover the soil.

f)   Leave the beakers until all bubbling has stopped.

g)   Add excess water to beakers B and D. Let them stand. The liquid should clear.

h)   Compare beakers A and B, then beakers C and D, and finally beakers B and D.

Discussion

Describe and explain your results. Why are beakers A and C necessary in this experiment? How might you speed up the reaction? Explain why one soil has more organic material than the other.

Organic material has other advantages to soil besides nutritional ones. One advantage can be demonstrated by lowering into water two wire baskets filled with two separate soil samples, one rich in organic content and the other mainly a mineral soil. Which soil sample disintegrates first? Why? When rain is falling on soil, what effect would organic matter in that soil have?

5.12   SOIL pH

pH is a measure of the concentration of hydrogen ions in a solution. It is the most common method of expressing acidity. pH is measured in a scale that runs from 0 to 14, with a pH of 7 representing a neutral solution. A solution with a pH below 7 is acidic. The more acidic the solution, the lower the *p*H. A solution with a *p*H above 7 is basic. The more basic the solution, the higher the *p*H.

Many minerals are more soluble in an acidic medium than in an alkaline one. As a result, the *p*H of a soil determines which minerals will remain in the upper soil layers and which will be leached through to the lower layers. This, in turn, will determine what organisms can survive in the soil.

Several methods for determining soil *p*H are available. A common and convenient method uses a *p*H indicator paper such as Fisher Alkacid Soil Test Ribbon. The Sudbury Soil Tester for Soil Acidity is also easy to use. Instructions on how to use these materials are in their packages. For more accurate results, the LaMotte soil kit is recommended.

Test several samples of soil, for example, topsoil and subsoil from a coniferous forest, an open field, and a deciduous forest.

Discussion

What kinds of soil tend to be acidic? Why? Were any of your samples basic? Why?

The usual *p*H range for soil is 4.0 to 8.5. Account for any *p*H values which were outside this range. Do the more acid soils appear to come from similar surroundings? Carry out the exercise in Section 5.10 on the same soils and compare the results.

5.13 TESTING FOR THE MAIN SOIL MINERALS

The elements nitrogen, phosphorus, potassium, and calcium are all essential to the growth and development of plants and animals. All of these, with the exception of nitrogen, are released from soil minerals by weathering. Their concentrations depend on the type of soil, the amount of precipitation, and the plant and animal life in the soil.

LaMotte Soil Testing Kits provide materials and instructions for the determination of the concentrations of these minerals in soil.

Test soils from various sites and depths. For example, determine the mineral content of the soil in a bog, in a clover patch, under a grove of deciduous trees, and in an area where plants do not grow well. Account for the results you obtain. Test soil from your home lawn and from the school lawn. What kind of fertilizer is being used on these lawns? Is it the correct kind? (A bag of fertilizer has a designation such as 7-10-5 on the front. This denotes, respectively, the percent nitrogen, phosphorus, and potassium in the fertilizer.)

Does a relationship exist between the minerals found in the soil and the soil pH? What is the annual precipitation in your region? How do you think it has affected your results?

5.14 RELEASE OF SOIL NUTRIENTS BY WEATHERING

Weathering is essential to both soil formation and nutrition of plants. Physical weathering breaks down soil particles without changing their chemical makeup. Chemical weathering changes the chemical composition of soil particles. As a result, new compounds are formed and nutrients are released into the soil.

In the following exercise, two methods for illustrating nutrient release are outlined. Method 1 involves the release of nutrients from a common soil mineral, apatite, which has the chemical formula $[Ca_3(PO_4)_2]_3 \cdot Ca(F,Cl)_2$. Method 2 involves the use of soil samples which are acted on by hydrochloric acid. In nature, weathering is assisted by carbonic acid which forms from carbon dioxide and water. In this exercise, hydrochloric acid is used in place of carbonic acid because the reactions take place faster.

Method 1 RELEASE OF NUTRIENTS FROM APATITE

Materials
a)   mortar and pestle
b)   piece of apatite approximately 3–6 mm in diameter
c)   250 ml beaker
d)   phenolphthalein solution
e)   Bunsen burner
f)   ring stand and ring
g)   funnel and filter paper
h)   test tube
i)   ammonium molybdate solution
j)   tin(II) chloride powder

Procedure

a)   Grind the apatite in the mortar and pestle.

b)   Place approximately 100 ml of distilled water in a beaker. Add the ground apatite.

c)   To this mixture add about 10 drops of phenolphthalein solution.

d)   Place the beaker on the ring stand. Heat to boiling.

e)   Record any changes observed.

f)   Filter about 5 ml of this mixture into the test tube.

g)   To the filtrate, add about 3 drops of ammonium molybdate solution.

h)   Shake the test tube.

i)   Add a few grains of tin(II) chloride powder to the test tube.

j)   Shake the test tube. Then let it stand for about 1 minute.

k)   Observe the color of the solution and record it.

Discussion

Phenolphthalein is colorless when neutral or acid and pink when in basic solution. If calcium were released from the apatite, it would make the solution basic. When you observed the solution before filtration, had calcium been released?

Phosphorus is present in the filtrate if it becomes blue after ammonium molybdate and tin(II) chloride have been added. Does apatite release phosphorus when it weathers?

Which parts of this exercise are closest to natural weathering? Why is the apatite content of soil usually low?

Method 2 RELEASE OF NUTRIENTS FROM SOIL

Materials

a)   2 Erlenmeyer flasks (250 ml)

b)   0.1M hydrochloric acid

c)   25 gm samples of 2 different soils

d)   Bunsen burner

e)   ring stand and ring

f)   2 beakers (250 ml)

g)   4 test tubes

h)   funnel and filter paper
i)   saturated ammonium oxalate solution
j)   ammonium molybdate solution
k)   tin(II) chloride powder

Procedure

a)   Label the Erlenmeyer flasks as A and B.
b)   Place one soil sample in flask A and the other in flask B.
c)   Into each flask place 100 ml of $0.1 M$ hydrochloric acid. Shake the flasks.
d)   Heat both flasks to the boiling point. Continue at this temperature for 15 to 20 minutes without allowing excessive boiling to occur.
e)   Filter the material in flask A into a beaker. From here transfer 5 ml of the filtrate into each of 2 test tubes and label them. Repeat the same procedure for flask B.
f)   To one test tube A and one test tube B, add 5 drops of saturated ammonium oxalate solution.
g)   Shake both tubes. Record any precipitates which form.
h)   To the other test tube A and test tube B, add 3 drops of ammonium molybdate solution and shake them.
i)   Add a few grains of tin(II) chloride powder to each test tube. Shake them.
j)   Let the test tubes stand for 1 minute.
k)   Observe the color in each test tube and record it.

Discussion

If calcium were released from either soil sample, a white precipitate would have formed in the test tube when the saturated ammonium oxalate was added. If phosphorus were present, the filtrate would have turned blue after ammonium molybdate and tin(II) chloride were added.

Do your two soil samples differ in the amounts of calcium and phosphorus released from them? Which soil would you expect to weather more quickly? Why?

5.15   SOIL FIXATION OF PHOSPHORUS

When phosphorus enters the soil in a soluble form (for example, as calcium phosphate), it is often rendered insoluble by a process called phosphorus fixation. In this insoluble form the phosphorus is not available to plants. Therefore it has no nutrient value. After a number of years, it may be rendered back into a usable form by changing soil conditions.

The fixation of phosphorus is affected by several factors. The pH of the soil; the presence of calcium carbonate, of iron, of aluminum, and of manganese in the soil; and the amount of clay in the soil, all affect how much phosphorus will stay in a soluble form.

In the following three methods, the influence of some of these factors is demonstrated.

A stock solution of calcium phosphate can be prepared by adding 0.2 gm of calcium phosphate to one liter of water which is then brought to the boiling point. The mixture is filtered and diluted 1:5 with distilled water. For these exercises all glassware should be well cleaned and then rinsed twice with distilled water. Why?

Method 1 THE EFFECT OF IRON

Materials

a) calcium phosphate solution
b) 2 test tubes
c) 1% solution of iron(II) sulfate
d) ammonium molybdate solution
e) tin(II) chloride powder
f) 1% solution of aluminum sulfate
g) 1% solution of magnesium sulfate

Procedure

a) To each test tube, add 5 ml of calcium phosphate solution.
b) Add 10 drops of 1% iron(II) sulfate solution to one of these test tubes. The other test tube serves as a control and should be labeled as such.
c) Into both test tubes, place 4 drops of ammonium molybdate solution.
d) Shake both test tubes.
e) Into each test tube, place a few grains of tin(II) chloride powder. Shake the tubes.

f)   Let the test tubes stand for 1 minute.

g)   Note any color in the 2 test tubes.

Discussion

Blue color in these test tubes indicates the presence of soluble phosphorus. The depth of the color indicates the amount present. A deep blue color indicates that much soluble phosphate is present. How does iron appear to affect the amount of soluble phosphorus in a solution? What effect would high concentrations of iron in soil have on phosphorus fixation? Repeat the experiment substituting, first, an aluminum compound for the iron, and then a manganese compound for the iron. Account for your results. How do these two elements affect phosphorus fixation?

Method 2 THE EFFECT OF *p*H
AND SOIL PARTICLE SIZE

Materials

a)   soil sample

b)   150 ml beakers

c)   calcium phosphate solution

d)   funnels and filter paper

e)   test tubes

f)   graduated cylinder

g)   ammonium molybdate solution

h)   tin(II) chloride powder

i)   soil *p*H kit

j)   soil sieves

Procedure

a)   Using the soil sieves, separate the soil sample into various parts, according to particle size.

b)   Test the *p*H of each part to insure that the *p*H is the same in each. (This should be so if the soil came from the same small region.)

c)   Place 25 gm of each soil part into a separate beaker. Label each one.

d)   To each beaker, add 40 ml of calcium phosphate solution. Swirl.

e)   With occasional additional swirling, let the beakers stand for 10 minutes.

f)   Filter each solution using separate funnels (or wash the funnel very well between filtrations). Place approximately 5 ml of the filtrate in a test tube.

g)   These separate test tubes should be labeled as the beakers were.

h)   Put approximately 5 ml of calcium phosphate solution in another test tube. Label it "Control."

i)   To every test tube, add 4 drops of ammonium molybdate solution. Shake the tubes.

j)   Into each test tube, place a few grains of tin(II) chloride powder. Shake the tubes. Let the tubes stand for 1 minute.

k)   Note the depth of color in each of the test tubes.

l)   Take a soil sample from another site where the soil appears to be of similar texture, but is at a different pH.

m)   Repeat steps a) to k) and then compare your results for particles of the same size.

Discussion

There is a large amount of soluble phosphorus in the control test tube. Note the intensity of the blue color. Color intensity indicates how much soluble phosphorus is present.

What is the relationship between the soil pH and the amount of soluble phosphorus in the soil? How does the soil texture appear to affect the amount of soluble phosphorus present? Account for this.

Imagine that you are on a field trip and that a soil sample has been taken. You have measured the pH and found the soil to be acidic. On examination the soil appears to have a clay texture. How much phosphorus fixation would you predict occurs in this soil? Why?

Method 3 THE EFFECT OF CALCIUM HYDROXIDE

Materials

a)   a soil sample of about 60 grams

b)   calcium hydroxide

c)   a solution which is $0.025M$ with respect to hydrochloric acid and $0.03M$ with respect to ammonium fluoride

d) 2 beakers (150 ml)
e) 2 funnels and filter paper
f) 2 test tubes
g) graduated cylinder
h) ammonium molybdate solution
i) tin(II) chloride powder
j) soil pH kit

Procedure

a) Separate the soil into 2 samples of approximately 30 gm each.

b) To one 30-gm sample add a pinch of calcium hydroxide. Mix it well.

c) Using approximately 5 gm from each soil sample, determine the pH of each (see Section 5.12). Record the value.

d) Place the remaining 25 gm of each sample in separate beakers. Label them.

e) Add 50 ml of the hydrochloric acid-ammonium fluoride solution to each beaker.

f) Swirl each beaker for 1 minute.

g) Filter each soil mixture. Place 5 ml of the filtrate in a test tube and label it.

h) Into each test tube, place 4 drops of ammonium molybdate solution.

i) Shake both test tubes.

j) To each test tube, add a few grains of tin(II) chloride powder. Shake the tubes.

k) Let the test tubes stand for 1 minute.

l) Note the color intensity in each test tube.

Discussion

The deeper the blue color in the test tube, the more soluble phosphorus is present.

How does the addition of calcium hydroxide affect the amount of soluble phosphorus in the soil? Is this related to the pH? Farmers frequently add calcium hydroxide to the soil. This is called "liming the soil." Why would this help to produce a better crop?

5.16 CARBON DIOXIDE PRODUCTION BY SOIL ORGANISMS

The process of respiration takes place in soil even though larger organisms, such as earthworms, may not be present. The details of the process involve many chemical equations, but it can be summarized generally by the following equation:

$$C_6H_{12}O_6 \ + \ 6\,O_2 \ \rightarrow \ 6\,CO_2 \ + \ 6\,H_2O \ + \ \text{Energy}$$

| Glucose | Oxygen | Carbon dioxide | Water | |

In order to determine how actively respiration is taking place in soil, the amount of carbon dioxide produced by that soil can be measured.

The following exercise measures the amount of carbon dioxide produced by a soil by determining how much carbon dioxide gas is absorbed by the alkali, sodium hydroxide. As sodium hydroxide absorbs more and more carbon dioxide, it becomes relatively more acidic. As a result, when much carbon dioxide is absorbed, less acid needs to be added in titration to reach the point where the indicator solution (in this case, phenolphthalein) changes from its alkaline color (pink) to its acid color (colorless).

Calculations can then be made to determine which soils produce the most carbon dioxide.

Materials

a) 3 wire loops fashioned from coat hangers
b) 3 crucibles
c) 3 Erlenmeyer flasks (250 ml)
d) 3 rubber stoppers
e) buret
f) 300 gm sample of soil
g) 0.5 gm glucose
h) $0.1M$ sodium hydroxide solution
i) phenolphthalein indicator solution
j) 50% solution of barium chloride
k) $0.1M$ hydrochloric acid

Procedure

a)   Divide the soil sample into three 100 gm portions.

b)   Using 100 gm of soil, calculate its moisture content and its water-holding capacity (see Sections 5.3 and 5.4).

c)   Leaving the control flask empty, add enough water to each of the other two flasks to bring 100 gm of original soil to 70% of its water-holding capacity.

d)   To one of these flasks add 100 gm of soil. To the other flask add 100 gm of soil mixed with 0.5 gm of glucose.

e)   Fix one wire loop into each of the three stoppers.

f)   Place 15 ml of $0.1M$ sodium hydroxide in each crucible. Suspend one in each loop.

g)   Moisten the rubber stoppers for an airtight seal. Place one in each flask (Fig. 5-8).

h)   Store all three flasks at $30°C$ for 24 hours.

i)   Open one flask at a time and remove the crucible. Add 1 ml of 50% barium chloride solution and 2 or 3 drops of phenolphthalein solution to the crucible.

j)   Slowly titrate the contents of the crucible with $0.1M$ hydrochloric acid, adding the acid drop by drop until the pink color disappears. Record the volume of hydrochloric acid used. (The contents of the crucible may be transferred to a clean Erlenmeyer flask to facilitate the titration. Rinse the crucible with distilled water and add the rinse water to the flask.)

k)   Repeat this titration with the contents of the other two crucibles. Record the volume of hydrochloric acid required for each.

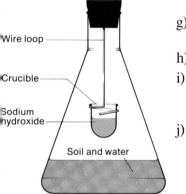

Wire loop

Crucible

Sodium hydroxide

Soil and water

Fig. 5-8
Measuring carbon dioxide production by organisms in soil.

Calculations

a)   (HCl, control, ml) $-$ (HCl, normal soil, ml) $\times$ 2.2

$$= \text{mg } CO_2 \text{ produced by 100 gm soil}$$

b)   Repeat this calculation for the soil supplemented with sugar.

Discussion

Why is a control flask necessary? What effect did sugar in the soil have on the amount of carbon dioxide produced by that soil?

Repeat this exercise with soils differing in organic content and explain your results.

Try the same experiment at different temperatures and account for any variations that occur.

5.17 COMPARING CARBON DIOXIDE CONTENT IN UNDISTURBED SOIL AND IN THE ATMOSPHERE

The principles involved in measuring the carbon dioxide production of soil are outlined in Section 5.16. However, in this exercise, limewater is used to show how much carbon dioxide is present. Limewater is calcium hydroxide solution. When carbon dioxide is bubbled through this solution, it becomes milky because a white precipitate, calcium carbonate, forms. If sufficient carbon dioxide is bubbled through the solution, it clears again because a soluble substance, calcium bicarbonate, forms.

Materials

a) metal pipe
b) rubber stoppers
c) 2 gas bottles
d) limewater
e) rubber tubing and glass tubing
f) screw clamps
g) large container with stoppered top and tap at base

Procedure

a) Set up the apparatus as shown in Figure 5-9.

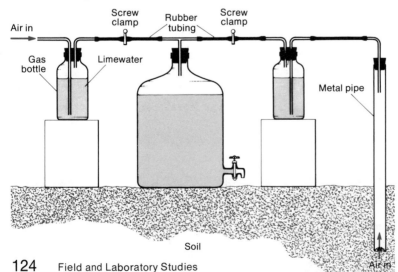

Fig. 5-9
Comparing the air in soil and in the atmosphere.

b)   Fill the two gas bottles with limewater.

c)   Close the clamps on the tubing leading to the center water container until it is filled with water.

d)   Open the clamps on the tubing. Then open the tap at the base of the water container. Use the clamps to equalize the air flow through the two gas bottles.

e)   If no change occurs in the appearance of the limewater, refill the water container and repeat steps c) and d) until a change does occur.

f)   Compare the clarity of the limewater in the two gas bottles.

Discussion

Why is there a difference in the carbon dioxide content of air and soil? Why is it necessary to have the clamps on the tubing? Perform this exercise inserting the pipe to different depths and in different soils. Account for any variations you find.

5.18   CARBON DIOXIDE PRODUCTION BY ROOTS

Much of the carbon dioxide found in soil is produced by organisms that reside in the soil. Yet, the roots of plants are living; therefore they must respire. It seems reasonable to assume, then, that they also add carbon dioxide to the soil. The purpose of this exercise is to check this assumption.

Materials

a)   3 healthy, growing seedlings (for example, lima bean)

b)   absorbent cotton

c)   3 test tubes

d)   bromthymol blue indicator solution

Procedure

a)   Put 10–15 ml of tap water in each of 3 test tubes. Add 3–4 drops of indicator solution.

b)   Insert the seedlings into the test tubes so that the roots are immersed in the dilute indicator solution.

c)   Wet the absorbent cotton and use it to support the seedlings.

d)	Observe the test tubes over the next 2–4 days. Record the results.

Discussion

Why is an indicator solution, which shows the presence or absence of acid, used to demonstrate the production of carbon dioxide?

Repeat this exercise using other types of seedlings. Does the indicator solution change color at a different rate? Why?

Does the size or nature of the root system appear to have any effect on the amount of carbon dioxide produced? Why?

5.19   THE PERMANENT WILTING PERCENTAGE

You have undoubtedly noticed that plants wilt when they are not given sufficient water. If plants are allowed to wilt excessively, they often fail to recover. In this exercise you can study the permanent wilting percentage for plants. This is the percentage of water in the soil below which a particular plant species will not recover from wilting.

Materials
a)	lettuce seeds
b)	pot of soil
c)	drying oven

Procedure
a)	Grow a lettuce plant in a pot of soil.
b)	Water the plant regularly until it is growing well.
c)	Allow the soil to dry until the plant wilts.
d)	Place a soil sample from the pot in an air-tight container.
e)	Rewater the soil to revive the plant.
f)	If the plant revives, return the soil to the pot.
g)	Repeat parts c), d), and e) until the plant fails to revive.
h)	Weigh the sample of soil and then determine the percentage of moisture in it (Section 5.3). This is the permanent wilting percentage for the lettuce plant.

Discussion

Your examination of the soil has shown you that a considerable amount of water present in the soil is not available to higher plants. In what forms do you think this soil water is found? Suppose you altered the microclimate of the lettuce plant. If you place one lettuce plant in direct sunlight and another in the shade, how would the wilting point be altered? Which plant would reach it first? What would be the effect of an electric fan blowing across the plant night and day? If you grew one lettuce plant in rich topsoil, and another in sandy soil, which would reach the wilting point first? Why? If you grew another species of plant (for example, radish) and carried through the experiment as for the lettuce, would you expect the permanent wilting percentage to be higher, lower, or the same? Try these experiments to check your predictions. On the basis of your studies, would you say that the permanent wilting percentage is primarily a characteristic of the plant or of the soil?

5.20 SOIL PREFERENCES OF CERTAIN PLANT SPECIES

You have probably noticed that certain species of plants seem to prefer certain soil conditions. For example, cacti grow best in soil with a low water and humus content; corn grows best in a heavily nitrogenated loam soil; black spruce grows in acid, water-logged bogs, but not in dry grassland areas.

In this exercise, you will choose a number of plants in your area and determine which soil conditions seem to promote their growth and development.

Materials

a) soil testing kit (for example, LaMotte)

b) materials for measuring organic content (see Section 5.10)

c) materials for measuring water content (see Section 5.3)

Procedure

a) Choose a site in the field where a particular plant species (for example, cedar trees) appears to be thriving.

b) Sample the soil throughout the area (about three times).

c)   Test each soil sample for potassium, nitrogen, phosphorus, water content, *p*H, and organic content. Record the results.

d)   Repeat steps a), b), and c) with three other plants. For example, choose a site where a patch of goldenrod is growing, another site where beech trees are common, and a third site where sphagnum moss is the dominant species. For meaningful results you should choose plant species which obviously prefer particular locations. Do not choose plant species that have been placed in particular locations by man.

Discussion

What particular conditions does each of your plant species prefer? In this exercise, you have measured only mineral content, organic content, *p*H, and moisture content of the soil. What other environmental factors help determine where particular plant species grow? What evidence is there that the plants may have helped to create the soil environment in which they are growing?

5.21   EFFECTS OF NUTRIENTS ON PLANT GROWTH

Plant growth is very dependent on the concentration of various minerals in the soil. Apparently, some minerals are essential to growth in height, whereas others are primarily involved in the tissue development of the plant.

In this exercise, you will add various minerals to the soil and observe their effects on the growth of selected plants.

Materials

a)   seedlings approximately 5 cm in height (for example, bean, corn, tomato)

b)   15 flowerpots, 8 cm in diameter

c)   a nitrogen compound, for example, sodium nitrate

d)   a phosphorus compound, for example, sodium phosphate

e)   a potassium compound, for example, potassium sulfate

f)   a commercial plant nutrient mixture which contains nitrogen, phosphorus, potassium, and more

g)   soil sample

h) evaporating dishes

i) drying oven

Procedure

a) Fill each of the flowerpots to 2 cm from the top with the soil sample.

b) Add 1 gm of sodium nitrate to each of 3 of the flowerpots. Mix it into the soil well. Label these 3 pots.

c) Add 1 gm of sodium phosphate to each of 3 other flowerpots and mix well. Label these pots.

d) Add 1 gm of potassium sulfate to each of 3 other pots. Mix well and label.

e) Add 2 gm of commercial plant nutrient mixture to each of 3 more pots. Mix well and label.

f) Label the remaining 3 pots as controls.

g) Transplant 1 seedling into each of the 15 flowerpots.

h) Water the plants regularly. Observe and record their height each day for at least 3 weeks.

i) At the end of this time, open the flowerpots. Gently remove the soil from around the roots. Record the extent of their growth. Note also any apparent deficiencies in stem and leaf structure.

j) Cut each plant into small pieces and place it in a weighed evaporating dish. (Label each dish before weighing.) Place the dish and its contents in a drying oven at 100°C for 24 hours. Weigh the dish and contents. Calculate the mass of dry plant material.

Discussion

Make graphs of the height of each plant versus time over the 3-week period. How does root development compare to the final height? Which mineral or minerals appear to promote growth in height and development of roots, leaves, and other plant parts? You may have to do further experiments to answer these questions completely. For example, you may find it necessary to:

a) Vary the concentration of sodium nitrate in a number of flowerpots to find out if there is an optimum concentration. Can too much sodium nitrate be added?

b) Determine if optimum concentrations of sodium phosphate and potassium sulfate exist.

c)   Determine the effectiveness of various combinations of the three minerals. For example, try the sodium nitrate and sodium phosphate together in one set of pots; the sodium phosphate and potassium sulfate in a second set; the sodium nitrate and potassium sulfate in a third set; and, finally, all three compounds together in a fourth set.

d)   Compare the effectiveness of various concentrations of commercial plant nutrient.

The concentration of each mineral supplied to the plant can be better controlled if water rather than soil is used as the supporting medium. This method of studying plant growth is known as "hydroponics." Information on hydroponics is available in several books. A good source is *A Sourcebook for the Biological Sciences*, 2nd ed., Morholt, Brandwein, and Joseph (Harcourt Brace Jovanovich, 1966), pp. 458-462. Pamphlets on hydroponics are probably also available from your Department of Agriculture.

The mineral content of soil may be one of the factors that determine the percentage of seeds in a given sample that germinate and produce healthy seedlings. Design and carry out an experiment to test this prediction. Do not forget to include suitable controls in your experimental design.

5.22   BURROWING HABITS IN GERBILS

You can readily imagine how impractical it would be to keep some mammals in the lab. One type which has recently become a pet in many homes in North America is the Asian gerbil, which just happens to be a soil burrower. The behavior of this animal can prove very interesting, especially when put in a semi-natural environment.

Whether or not you can set up the following display depends on the particular nature of your laboratory.

Materials

a)   available and appropriate counter-top space

b)   clean sand

c)   aquarium gauge glass to completely enclose the counter-top area as shown in Figure 5-10.

d)   cloth tape

e)   male and female gerbils

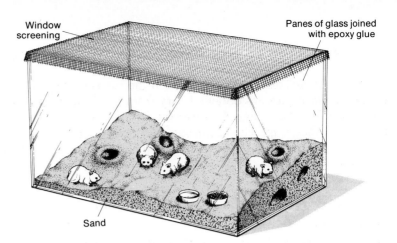

Fig. 5-10
A cage that permits close study of a small gerbil colony.

Window screening

Panes of glass joined with epoxy glue

Sand

f)   wood, nails, and epoxy glue to construct framework to hold glass

Procedure

a)   Build the glass cage. If space permits, make the cage 50 cm tall and have it enclose an area that is 50 cm by 100 cm.

b)   Place about 25 cm of clean sand on the counter-top.

c)   Let the gerbils do the rest. Just feed and water them as prescribed by the pet store.

d)   Do not let the colony get overcrowded. Gerbils can get very vicious with one another under such circumstances.

e)   Observe and record the time spent burrowing, the uses of burrows, the differences in the roles of adult males and females, and the development of the young to adulthood.

Discussion

In what ways have gerbils adapted to a life in the soil environment? What useful functions do gerbils perform in the soil? Do they have any harmful effects on the soil?

Would your gerbil colony survive if it was released in the vicinity of your school? Why?

Many experts claim that imported animals like the gerbil could affect the ecology of some localities if enough of them escaped from captivity. Where in North America would gerbils constitute the greatest threat to the ecology? Why?

Do not release the gerbils at the end of the study. Pet stores will take the ones that are not claimed as pets by members of the class.

5.23 SAMPLING SOIL ORGANISMS AT THE SURFACE

Many soil organisms can be collected at the surface using simple methods. The organisms obtained vary from hard-shelled arthropods to soft-bodied slugs. In this exercise, you will set up two types of surface trap. The organisms will be attracted to one or the other depending on the environmental conditions they prefer.

Materials

a) several pieces of plate glass (30 × 20 cm)
b) resin
c) several unpainted boards (2 × 30 × 20 cm)

Procedure

a) Paint the plate glass pieces with resin on one side.
b) When the resin is sticky, leave the glass plates with the sticky side uppermost at various sites (for example, in a coniferous woods, in a grassy area, and on garden soil).
c) After 48 hours, identify the organisms caught and record your results.
d) Place the boards at approximately the same sites as the glass plates. Leave them for 2 to 3 weeks.
e) At the end of this time, turn the boards over. Identify the organisms beneath the boards and record your results.

Discussion

What environmental conditions does each organism seem to prefer? How does the structure of each organism match its habitat choice (for example, the length of its legs, the presence or absence of an exoskeleton, and the type of mouthparts)?

Repeat this procedure at the same sites during different seasons, and during wet and dry periods. Account for any variations that you notice.

5.24 RESPONSES OF EARTHWORMS

It is easy to see that earthworms are more plentiful in some soils than in others. But the reasons why are not as readily seen. In the following exercise, various soil conditions are reproduced in the laboratory in order to observe how earthworms respond to them.

Materials

a) shallow pans (like small dissecting trays)

b) 2 soil samples, one of sandy soil and the other of soil rich in organic material

c) several earthworms

d) dark paper

e) light source

f) dilute hydrochloric acid (0.1M or less)

g) calcium hydroxide solution (limewater)

Procedure

a) Spread rich organic soil to a depth of 5 cm over the bottom of one of the pans.

b) Cover one-half of the length of the pan with dark paper. Place a light source above the pan.

c) Place 2 earthworms at the center of the pan on top of the soil. Record your observations.

d) Remove the earthworms from the soil. Then oven-dry the soil (see Section 5.3).

e) Spread the dried soil evenly in the pan. Add enough water at one end that approximately one-half the length of the pan becomes moist.

f) Place 2 earthworms at the leading edge of the water. Record your observations.

g) Remove the earthworms and moisten all of the soil in the pan. Divide this soil into two equal parts.

h) Determine the pH of the moist soil (see Section 5.12).

i) Treat one-half of the soil so that it has a pH above 4.5, and the other half so that it has a pH below 4.5. (You will have to decide in each case whether to use hydrochloric acid or limewater.)

j) Fill one end of the pan with the soil of the lower pH and the other end of the pan with the soil of the higher pH.

k) Place 2 earthworms at the point where these two soils meet. Record your observations.

l) Using another pan, fill one-half the length of the pan with rich organic soil. Fill the remainder with the sandy soil. Keep the soil evenly moist. Add two earthworms. Observe after one week and record your results.

Discussion

What soil conditions does your earthworm species appear to prefer? If you were going to dig for worms for bait, what soil conditions would you look for? How could you force the earthworms to come to the surface?

Earthworms can be taught to run a very simple maze. The usual one is a T-shaped maze, with rich organic soil at one end of the crossbar and either dilute acid on filter paper or sandpaper at the opposite end of the crossbar. If you wish to try it, use only hungry earthworms. Mark them with water insoluble ink so that you can keep track of their trials-and-errors. Finally, remember that earthworms are slow learners and it may take 50 or more runs of the maze to train them!

5.25 FIELD SOIL CONDITIONS AND EARTHWORM NUMBERS

In Section 5.24, you determined some of the soil conditions which earthworms prefer. You can verify your discoveries in the field by counting the numbers of earthworms in soils which vary in such factors as pH, organic content, and moisture content. The following exercise outlines a method for doing this.

Materials

a) potassium permanganate solution, 1.5 gm of potassium permanganate per liter of water

b) covered containers partially filled with moist peat moss

Procedure

a) Choose a site and mark off a 1 meter square plot.

b) Remove all of the surface vegetation within the square, but leave the roots so that the area can regrow.

c) Pour about 7 liters of solution evenly into the marked-off area. Allow time for it to soak in.

d) As the earthworms come to the surface, count them and note their species, if possible. Place them in the covered containers. Continue to remove the earthworms until no more come to the surface. Do not dig for the worms.

e) Repeat this procedure with three more 1 meter square plots in the immediate area. As a result, you should have four completed earthworm counts from the same soil type.

f)   Average the four earthworm counts.

g)   Take a soil sample from this area. Test its pH (Section 5.12), moisture content (Section 5.3), temperature (Section 5.2), and organic content (Section 5.10).

h)   Repeat this entire procedure in two other areas where soil types are different.

i)   Record all of your results in chart form, under the following column headings: site, numbers of each earthworm species, pH, percent moisture content, temperature, and percent organic content.

Discussion

Outline what soil conditions each species of earthworm seems to prefer. Do these results agree with those found in Section 5.24? Account for any differences. Why is it an advantage to have earthworms in a soil? Does any relationship exist between the size of the earthworm population in an area and the plant life growing there? There are some sources of error in this exercise. What are they?

Cautionary Note

Fig. 5-11
Studying earthworm activity. Place the glass plates 1 cm apart. Plates about 30–40 cm square are most easily handled.

This exercise should not be carried out on property without the owner's express permission. Larger properties should be used so that removal of several square meters of surface vegetation will not be too noticeable.

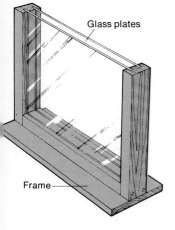

Glass plates

Frame

5.26   EARTHWORMS AND SOIL MODIFICATION

Earthworm activity in the soil can increase the total volume of some layers by as much as 20%. In this experiment, you will investigate this effect and other types of behavior such as tunneling, forming casts (the waste material piled at the entrance of a burrow), soil mixing, and the collecting, burial, and digestion of surface plant litter.

Materials

a)   2 glass sheets of equal size

b)   wooden frame (see Figure 5-11)

c) one or more earthworms

d) 3 or more soil samples (preferably of different colors, one rich in organic material, and at least one made up of fine clay).

e) a few partly decayed leaves (not oak)

f) small piece of wire screen and weights

Procedure

a) Set up the apparatus as shown in Figure 5-11.

b) Make sure that all soils are just slightly moist.

c) Place a layer of soil 5-10 cm deep evenly across the bottom of the apparatus.

d) Place a similar layer of the clay soil evenly on top of the first layer.

e) Continue to add each different soil type in turn. Try to have contrasting colors next to one another so you will be able to identify the layers later.

f) Place a 5 cm layer of organically rich soil across the surface. Leave a 10 cm air space at the top of the glass panels.

g) Draw a line on the glass to show the upper level of the top layer.

h) Place a few leaf fragments on the surface.

i) Place a healthy worm in the apparatus.

j) Place a wire screen on top of the apparatus. Hold it in place by weights to prevent the worm from escaping.

k) Observe the activity of the earthworm over a period of 10–20 minutes. Record its behavior.

l) Place the apparatus in a darkened area for 24 hours.

m) Observe again, marking any change in the upper level of the soil.

n) Calculate any change in the total soil volume.

o) Trace the total length of tunnels created in 24 hours.

p) Calculate the amount of tunneling that would occur in 1 year at the rate calculated for the first 24 hours.

q) Note the displacement of any soil from one layer to another.

r) Compare the amount of tunneling done in each layer.

s) Observe any changes in the leaf litter at the surface.

t) Continue the experiment if desired, keeping the soil slightly moist and the apparatus in darkness.

Discussion

Why is the displacement of soil from one layer to another important? Do you think there is some maximum to the increase in volume that worms can cause? Will an increase in volume occur more likely in upper or lower layers? Why?

5.27 SEPARATING SOIL ARTHROPODS FROM LITTER AND SOIL OF HIGH ORGANIC CONTENT

Of all the small soil life, arthropods are perhaps the easiest to remove and examine. One of the original pieces of apparatus for doing this is called a Berlese funnel. A variation on this, which you will use, is called a Tullgren funnel. Although not very satisfactory for isolating arthropods from soils with a high clay content, this apparatus is otherwise effective and easy to construct.

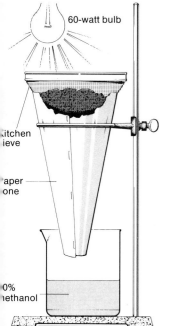

60-watt bulb

kitchen sieve

paper cone

70% methanol

Fig. 5-12
A Tullgren funnel for isolating arthropods from soil.

Materials

a) 60-watt light bulb with a shade
b) sieve
c) ring stand and ring
d) 250 ml beaker
e) heavy paper cone with tip removed
f) 70% methanol
g) 2 soil samples (topsoil including leaf litter and subsoil)
h) microscope and accessories

Procedure

a) Be sure that the two soil samples are kept moist until the time of the exercise.
b) Assemble the apparatus as in Figure 5-12.
c) Fill the sieve about half-full with the topsoil and leaf litter sample. Be sure there is a space between the soil and the funnel so that air can circulate around the sample.

d) Fill the beaker half-full of 70% methanol. This acts as a preservative for the arthropods leaving the soil.

e) Turn on the light. Be sure it is 15–20 cm away from the soil surface. If it is too close, the soil will heat too quickly and some of the animals will die within the soil. As a result, your count will not be accurate.

f) After 24 hours, count and identify the arthropods (see Section 3.4), using a microscope.

g) Remove the soil from the sieve. Determine the percent organic content in the soil sample (see Section 5.10). Record your results.

h) Repeat this whole procedure using the sample of subsoil.

Discussion

How is this funnel able to separate the arthropods from the soil? What does this indicate about the living conditions which soil arthropods prefer?

How is the amount of organic material in a soil related to its arthropod population?

Attempt this procedure in the winter, particularly in marsh or bog areas, by obtaining frozen samples, bringing them to room temperature, and then separating out the various organisms.

5.28 FLOTATION EXTRACTION OF ARTHROPODS FROM HUMUS AND MINERAL SOILS

The Tullgren method for arthropod extraction (Section 5.27) is useful for organisms capable of easy movement through loose soil or litter material. Many organisms are very tiny and live in the more compact humus and mineral soil layers. They often are unable to escape through the soil, which is a necessary feature if the Tullgren funnel is to work. Fortunately, because they are less dense than water, most soil arthropods can be separated by flotation. Flotation gives fairly quick results, although as with any method, many organisms will fail to be collected.

Materials

a) trowel

b) plastic sandwich bags with ties

c) water bucket and water

d) salt

e) several petri dishes

f) dissecting microscope

g) aquarium air pump

h) rubber tubing and aquarium aerator stone

i) beaker

j) filter paper to fit petri dish

Procedure

a) Collect soil samples that half fill the plastic bags. Place label tags on the samples, and tie the tops.

b) Fill a bucket with water within 2 cm of the top. Add a cup of salt and stir until most of it has dissolved.

c) Before opening the bags, roll the soil between thumb and forefinger to break up large earth clumps into smaller particles. Do not do this too vigorously (you might destroy some of the invertebrates).

d) Create turbulence and a great many air bubbles in the water using an aquarium pump, tubing, and aerator stone. *Slowly* pour the soil into the bucket where the greatest number of bubbles are coming to the surface.

e) Skim off the surface water into a beaker to collect the floating material.

f) Place a piece of filter paper in a petri dish. Pour 2–3 mm of the skimmed water over the bottom. Repeat, using additional petri dishes, and pouring only 2–3 mm of water at a time.

g) Observe under a dissecting microscope. Try not to disturb the water in any way while looking for organisms. They are small, but will give away their presence by moving.

h) Organisms collected can be stored in 60–90% ethyl alcohol.

i) This procedure can be performed in the field using an innertube filled with air as a source of air for the aerator stone. However, a reasonable separation can be obtained without any source of air.

Discussion

Identify and account for the presence of the various soil organisms isolated by this method.

Many modifications of this method have been used by scientists. Improve on the method in whatever ways you can.

5.29 RESPONSES OF SOIL ARTHROPODS

The easiest way to study the responses of some of the larger soil arthropods is to set up a colony of them in the classroom. Here, their responses to light, moisture, and soil types can be seen. Of added benefit is the observation of the social behavior of these interesting creatures.

In the following exercise, instructions are given for setting up ant and termite colonies. With suitable modifications, any large soil arthropods can be studied in this way.

Materials

a) large glass containers (preferably small aquaria)
b) dark paper
c) sandy soil and soil rich in organic material
d) petri dishes
e) balsa wood
f) filter paper
g) sponge
h) ants, termites, and other soil arthropods

Procedure

a) Fill one-half the length of the aquarium with sandy soil and one-half with soil rich in organic material.

b) Tape dark paper to the sides and ends of the container. Secure the paper in such a way that it can be lifted at intervals to observe the responses to light and the burrowing patterns.

c) Obtain a small ant hill from a field, or select some ants from the field. Be sure to have some workers and a queen in those ants selected.

d) In order to prevent the ants from escaping, one of two things can be done. The top can be sealed with a porous material that allows gaseous exchange (for example, vinyl screening), or the container in which the ants are kept can be placed in a basin of water to form a moat.

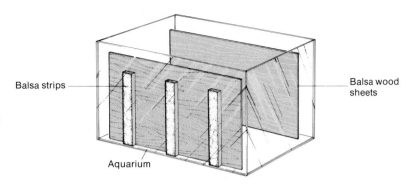

Fig. 5-13
Apparatus for a classroom
termite colony.

Balsa strips

Balsa wood
sheets

Aquarium

e) From time to time, feed the ants bits of raw potato, milk-soaked bread crumbs, lettuce, dilute molasses or honey, and dead insects. Be sure to remove any excess food so that it will not become moldy.

f) After a few days, note which part of the container (sandy or organic) the ant colony is in. Also note the ants' response to light.

g) Occasionally moisten a small piece of sponge kept at one end of the container. Note the ants' response to moisture.

h) The reactions of an established colony toward intruders can be observed by introducing members of other ant species or other invertebrates into the aquarium.

i) To start a termite colony, begin with a flat-walled glass container like an aquarium. Place a piece of balsa wood against the inside of each of the two long walls (Fig. 5-13).

j) So the termites can move easily between the balsa wood and the glass wall, place small strips of balsa wood between them. Secure dark paper to the outside of the glass.

k) Fill the container about one-quarter full of soil, preferably soil rich in organic material.

l) Find a termite colony in an old rotting log. Move as much of it as possible back to the classroom.

m) Gently remove the termites from the log fragments. Transfer them to the container you have prepared. Do not allow termites to escape into the classroom!

n) Observe their responses to light and various foods which are added (different types of vegetation, bread, and so on).

o) Go into the field and observe the field conditions of other soil arthropods such as centipedes, millipedes, and wood lice. Duplicate these field conditions as closely as possible in the laboratory and set up your own colonies of these organisms. Design experiments to study their habitat preferences.

A Useful Gadget. Small arthropods, like ants, can be collected and transported simply and untouched by human hands with an aspirator. It acts like a little vacuum cleaner with you supplying the suction. Figure 5-14 shows one version of this gadget. Be sure that the cloth screen is firmly in place at all times. Ants don't taste very good.

Discussion

What type of soil conditions (light, moisture, organic content) do each of the arthropods studied appear to prefer? What adaptations do they possess that suit them to their habitats? What niches do they occupy in food chains? How are they adapted to fill these niches?

What advantages does colonization give to these organisms? What are the disadvantages?

5.30 PASSAGE AND CHAMBER BUILDING BY ANTS

Using an aquarium to study ants (Section 5.29) is good for keeping a permanent colony, but is poor for observing subsurface activities. The apparatus described in Section 5.26 for earthworms can be used for ants with extremely interesting results. Ants will avoid tunneling and underground chamber construction up against the sides of the glass unless forced to do so. This apparatus, unlike an aquarium, forces them to build within your view, whether they like it or not. Some 40–50 ants with an equal number of larvae and pupae, will give you real insight into ant life. The ants will build with or without a queen, so only a part of an ant colony is necessary in this experiment.

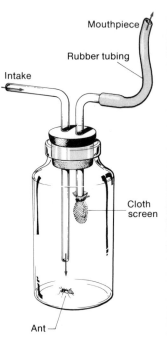

Fig. 5-14
An aspirator like this is useful for collecting small arthropods.

Materials

a)   2 glass sheets of equal size

b)   wooden frame (see Figure 5-11)

c)   40–50 ant adults and 40–50 larvae and pupae

d)   soil, not too fine in texture but loose and easily worked by ants

e)   tray and water

f)   2 sheets of black paper and tape

g)   marking pen for glass

Procedure

a)   Set up the apparatus as shown in Figure 5-11.

b)   A grid should be made on the glass, using a marking pen. Make the grid large enough to cover the complete area in which you plan to put soil. Make the grid lines 2 cm apart. Number them from top to bottom and left to right.

c)   Fill the soil chamber $\frac{1}{2} - \frac{2}{3}$ full of soil.

d)   Add the soil from the ant colony containing adults, pupae, and larvae, preferably so that the final surface corresponds with the "0" line of the grid.

e)   Place the apparatus in a tray containing water so that a moat at least 20 cm in width surrounds the apparatus.

f)   After observing initial behavior for 10–15 minutes, tape black paper to each side of the container to let the ants work under normal dark conditions.

g)   Observe the ants 24 hours later.

h)   Note the position of various chambers in a table in your notes. Record the apparent use, if any, of each chamber. In chambers containing pupae or larvae, count the number present. Record in a separate column of your table any other miscellaneous information you think might be important.

i)   Each 24 hours examine the apparatus again. Record the appearance of new chambers, their use, numbers of larvae or pupae, and so on.

j)   Note changes in the uses of old chambers.

k)   To vary conditions and observe specific responses of the ants, try steps l) through r), one at a time, at various intervals in this study. Wait 24 hours to observe responses to changes in conditions. Record your results.

l)   Cut a "porthole" in the black paper to let light into a particular chamber.

m)   Add water to one end of the apparatus to dampen the soil 10-12 cm deep.

n)   Place food, such as syrup, in a small container on the surface. Observe after 30–60 minutes for best results.

o)   Shine a strong light on the surface.

p)   Put a rich organic mulch 2 cm in thickness on the top.

q)   Add 10–15 "foreign" ants on the surface.

r)   Add an earthworm or other soil inhabitant.

Discussion

"The activity of each member ant in an ant society is ultimately for the perpetuation of the society." List each activity that you observed and describe how you think it aids in the success of the colony.

5.31   REACTIONS OF ARTHROPODS TO DIFFERENT SURFACES

Most soil organisms spend much of their lives in darkness. Of those that do come to the surface, many do so only at night. For these animals, good eyesight is not nearly as important as a good sense of touch. The wood louse is an example of an organism that lives primarily in darkness and relies heavily on its ability to distinguish between desirable and undesirable conditions by means of its sense of touch. This experiment is designed to show the reactions of wood lice to different surfaces.

Materials

a)   2 or 3 adult wood lice (0.5–1 cm long)

b)   culture bottle containing moist soil

c)   1–2 petri dishes

d)   paper and pencil

e)   stop watch or watch with second hand

f)   scissors

g)   pieces of glazed or waxed paper, unglazed paper, blotting paper, silk, wool, flannel, other fabrics

h)   cellophane tape or cloth tape

Procedure

a)   Place the bottom section of a petri dish on one of the materials selected for study.

b)   Trace the dish outline with a pencil, and then cut out this circular section.

c)   Fold the circular section in half, unfold, and cut along the fold to get two identical half sections.

d)   Do this with the remaining types of material.

e)   Tape together, on the bottom surfaces, those materials you wish to compare (two at a time).

f)   Place the petri dish lid on a desk surface.

g)   Place the first test surfaces in the petri dish lid so that the tape side is down.

h)   Make a circular tracing of the dish bottom on a piece of white paper.

i)   Remove one wood louse from the culture bottle. Place it in the center of the petri dish lid.

j)   Place the bottom in place within the sides of the lid.

k)   With a pencil, draw on the white paper the path taken by the louse over a period of 5 minutes.

l)   Make sure to note on the white paper which half was which material.

m)   Repeat using different combinations of material.

n)   Use other wood lice to repeat your observations, if time permits.

Discussion

Make a chart showing in sequence the surfaces most preferred as opposed to those least preferred. Do wood lice have the ability to sense differences in the smoothness of surfaces?

Why do wood lice respond as they do? Is the sense of touch necessarily the only factor governing their behavior in this experiment?

Select other organisms which could be tested using this apparatus. Predict their behavior and test your predictions.

In what ways might the experiment be improved?

5.32   MICROSCOPIC EXAMINATION OF ARTHROPODS

Because of the small size of most soil arthropods, it is often necessary to use a microscope to study them. Some people are interested only in quick identification, while others may wish to study specimens in detail. Procedures are given to meet the needs of both groups. You can decide how much effort you wish to put into slide preparation.

A. PRESERVATION

Arthropods collected in the field should be kept in small vials and preserved in either 70–85% ethyl alcohol or Oudeman's fluid. The latter is made up of 87 parts by volume of 70% alcohol, 5 parts by volume of glycerol, and 8 parts by volume of glacial acetic acid. The glycerol prevents stored material from becoming brittle.

B. REMOVING INTERNAL CONTENTS

This step is well worth the effort if time is available. By placing the specimen in certain solutions, the soft internal tissues are destroyed, leaving the hard outer skeleton intact. The specimen becomes softened and is much more easily viewed on a slide. Two procedures are described.

Materials

a) caustic potash (10% potassium hydroxide in water)

b) lactic acid, 50–100%

c) test tubes

d) Bunsen burner

e) sand

f) eye dropper

Procedure

Caustic Potash

a) Caustic potash is a strong dissolving agent and must be handled with care.

b) Small specimens can often be treated in cold potash overnight.

c) Larger specimens can be boiled in this fluid for 5 minutes. To prevent splashing over the sides of a test tube, place a few grains of dry sand on the bottom. Apply the Bunsen burner flame gently to the side of the test tube near the top of the 5 cm of liquid in it. Heating within a water bath (putting the tubes in a glass beaker containing heated water) is a safer technique. This also permits many specimens to be prepared at once.

d) Do not heat too long or too much of the specimen will disappear. A small puncture in the abdomen of larger specimens will speed up the process.

e) When the internal contents are completely gone, draw off the potash with an eye dropper. Replace it with clean water. Repeat several times.

Lactic Acid

a) Specimens with weak exoskeletons should be treated with 50–100% lactic acid instead of caustic potash. The weaker the skeleton, the lower should be the concentration of lactic acid.

b) Tiny specimens can be treated in a few drops of solution on a slide. Gently warm the slide on a hot plate until the specimen loses its internal parts.

C. MOUNTING

Materials

a) depression slides
b) cover slips
c) alcohol or Oudeman's fluid
d) Canada balsam
e) glacial acetic acid
f) clove oil or cedar-wood oil
g) xylol

Procedure

Non-permanent Slides

a) Specimens which will eventually be returned to the original preserving solution should be mounted in water or the preservative on a depression slide. (This type of slide has a small cavity on the surface into which the organism and fluid can be placed.)

b) A cover slip placed over the cavity should prevent evaporation of the fluid.

Permanent Slides

a) Place the specimen, with internal structures removed, into glacial acetic acid for 5 minutes to remove any water.

b) Then place the specimen in clove oil or cedar-wood oil. Patches of milky color will appear if all of the water has not been removed. If this happens, the specimen should be returned for a short while to the acetic acid.

c) After 5 minutes in the oil, place the specimen on a slide. Blot most of the oil away from around the organism.

d) Place a drop of Canada balsam on the mount. (This is a resin and is a fairly good permanent mounting substance.) The balsam should be just able to run. That is, it should not be too tacky nor too fluid. Dilute it with a small drop of xylol if it is too tacky.

e) Lower a cover slip onto the balsam, using a pin to support one edge as the cover slip slowly settles into place.

f) The balsam requires several weeks to harden at room temperature. Keep the slide in a horizontal position at all times until the balsam is hard. To speed up the hardening, use a slight amount of heat from something like a light bulb.

Discussion

Whenever examining specimens under a microscope, keep two things in mind, *structure* and *function*. They go hand in hand. Every part of an organism's anatomy has a function important in its life processes. Often it is difficult to determine how a particular structure might function, especially when the animal is dead. Keep in mind where the animal lives, what it eats, what enemies it has, and so on. Then you can probably make reasonable, educated guesses.

If you wish to draw diagrams of your specimens, use the drawings of this book as a guide to technique. Try to give a description of the function of any structures drawn. If you can include brief statements of functions right on your drawing, you will make it much more worthwhile and informative. Pay particular attention to the ways in which your specimens are suited to a life in the soil environment.

5.33 ENCHYTRAEIDS: COLLECTING, CULTURING, AND EXAMINING

Moist leaf litter often contains a large number of tiny "white worms." One of the best ways to remove these organisms from the soil is by the *wet sieving* technique described here. Once collected, colonies of these worms can be established quite easily.

They can be used as food for a number of laboratory animals including salamanders, crayfish, guppies, and other fishes. Also, because of their size and transparent flesh, they are useful animals for the demonstration of worm-like motion and feeding under the microscope. Food can even be observed moving through the intestine.

A. COLLECTION

Materials

a) moist, partly decomposed leaf litter
b) coarse and fine mesh screening
c) container similar to that shown in Figure 5-15
d) fine cloth

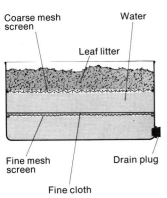

Fig. 5-15
Apparatus used to collect enchytraeids.

Procedure

a) Set up the apparatus as shown in Figure 5-15.
b) Place the litter on the coarse screen. Gently stir and agitate the material to separate the invertebrates from the litter.
c) Let the apparatus sit for 15–30 minutes. Then drain off the water.
d) The "white worms" should get caught in the fine cloth from which they can then be carefully removed.

B. CULTURING

Materials

a) plastic or wooden containers with lids
b) rich, dark, moist soil
c) bread
d) milk

Procedure

a) Place 5–10 cm of the moist soil in a container. (Many small colonies are better than one large one, in case a culture fails to multiply for some reason.)
b) Place 10–20 worms in the soil to "seed" it.
c) One centimeter squares of bread soaked in milk can be buried at a few sites in the container (the number varying according to the size of the container).

d)   Replenish the food when necessary.

e)   Keep the cultures loosely covered with a lid, but allow free access of air.

C. MICROSCOPIC EXAMINATION

Materials

standard microscopic equipment

Procedure

a)   Place live specimens in a drop of water on a slide.

b)   Before placing a cover slip on the slide, be sure a small amount of soil or a few sand grains are in the water. This will prevent the worms from being crushed.

Discussion

Make a summary of the environmental conditions that seem to be preferred by enchytraeids. From your microscopic examination of these animals, suggest reasons why enchytraeids prefer these conditions.

Do you think that the environmental conditions determine the nature of the enchytraeids, or that the enchytraeids select a particular environment because of their nature?

5.34   SEPARATING NEMATODES FROM SOIL SAMPLES

Nematodes live in the water which surrounds the soil particles. These worms appear as very fine threads to the naked eye, and are difficult to separate from soil particles. One method for doing so uses a Baermann funnel. This method is described.

Materials

a)   2 large glass funnels

b)   2 light bulbs (25 watt)

c)   rubber tubing

d)   2 pinch clamps

e)   2 beakers (250 ml)

f)   2 ring stands and rings

g)   2 glass rods

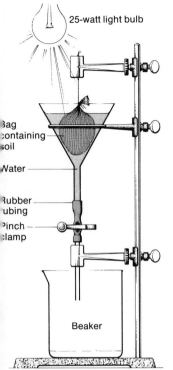

25-watt light bulb

Bag
containing
soil

Water

Rubber
tubing

Pinch
clamp

Beaker

Fig. 5-16
A Baermann funnel for separating nematodes from soil.

h)   cheesecloth

i)   equal size samples of topsoil and subsoil

j)   microscope

k)   glass slides and cover slips

Procedure

a)   Enclose each soil sample in a bag of cheesecloth, tightly securing the top.

b)   Assemble the apparatus as shown in Figure 5-16 for each soil sample.

c)   Be sure that the cheesecloth bag is submerged in water. When suspending the bag from the glass rod, check that there is a space between the wall of the funnel and the bag.

d)   Place the light source about 10 cm from the surface of the water. Turn on the light and leave it for about 24 hours.

e)   Open the clamp and allow the water (and the organisms in it) to empty into the beaker.

f)   Mount some of the organisms in water on a glass slide. Observe and identify them under the microscope.

g)   Count the numbers of nematodes present in each sample, either directly or by making dilutions of the contents of the beaker.

Discussion

This separation is possible because of the preference of soil nematodes for particular soil conditions. What are they? How does the number of soil nematodes relate to the organic content of the soil?

Repeat this procedure comparing the population density of nematodes in the upper layers of two different soils. Compare the organic content as well (see Section 5.10).

Attempt this separation in two different seasons in the same soil. How does temperature affect the nematode population?

5.35   SOIL PROTOZOA

Protozoa live in the moisture surrounding soil particles. They are quite difficult to culture outside of soil because they require special media. In these media protozoa thrive, but other microorganisms do not. The following exercise allows you to separate and culture several types of protozoa, but your population counts will not be as accurate as they would be with more complex methods.

Materials

a) 1,050 gm of soil rich in organic material
b) sterile water
c) test tubes
d) 1,000 ml and 2,000 ml Erlenmeyer flasks
e) autoclave
f) calcium carbonate
g) funnel and filter paper
h) dibasic potassium phosphate (K_2HPO_4)
i) cotton plugs
j) inoculating loop
k) microscope
l) microscope slides
m) 50 gm of sandy soil

Procedure

a) Place 1,000 gm of soil in 1 liter of tap water. Use a 2,000 ml flask.
b) Autoclave this suspension at 5 to 10 lb pressure for 30 minutes.
c) Add 0.5 gm of calcium carbonate to the suspension.
d) Filter it repeatedly until it is clear. This is the soil extract solution.
e) Combine 100 ml of soil extract solution, 0.5 gm dibasic potassium phosphate, and 900 ml of tap water in a flask.
f) Into each of 8 test tubes, place 10 ml of this solution. Stopper the test tubes with cotton plugs.
g) Sterilize these test tubes in the autoclave. This is the medium for the protozoa.
h) To 500 ml of sterile water in a 1,000 ml Erlenmeyer flask, add 50 gm of soil. Shake it vigorously for 5 minutes.

i) Allow the coarse particles to settle.

j) Dilute this soil suspension in the following way:
1) Add 1 ml of the suspension to 9 ml of sterile water in a test tube. This is a 1:100 dilution.
2) Add 1 ml of dilution 1) to 9 ml of sterile water. This is a 1:1,000 dilution.
3) Add 1 ml of dilution 2) to 9 ml of sterile water. This is a 1:10,000 dilution.
4) Add 1 ml of dilution 3) to 9 ml of sterile water. This is a 1:100,000 dilution.

k) Into 2 of the sterilized test tubes of protozoa medium, add 1 ml of dilution 1). Repeat for dilutions 2), 3), and 4). Label the test tubes.

l) Leave these test tubes at room temperature.

m) Every 2 days transfer a small amount of each culture to a microscope slide. Use an inoculating loop. Examine each slide under the microscope.

n) Identify the species of protozoa and record them. Note the number of protozoa taken in one loopful of each culture.

o) Repeat steps f) through n) with a more sandy soil having approximately the same moisture content. Record the results.

Discussion

In which type of soil are protozoa more plentiful? How would you expect to find them distributed through topsoil and subsoil? Why?

Repeat steps f) through n) on soils which have approximately the same organic content, but which differ in their usual moisture content. Do you wish to revise your answers to the first two questions of the discussion?

5.36 BACTERIA FROM SOIL

The number of bacteria in any soil sample varies with the amount of moisture and the amount of organic material in that soil. Most exercises designed to estimate bacterial populations require tedious media preparation. The Millipore method outlined here is less time-consuming and yet very effective.

Materials

a)   soil samples (one from topsoil, one from subsoil)

b)   Millipore kit to include:
      Swinnex-25 filter holder
      Type GS filter
      6 small dilution tubes
      dilution tube rack
      Sterifil apparatus
      Type HAWG filter
      dilution loop
      sterile petri dishes
      Total Count Medium
      plastic syringe
      forceps

c)   distilled water (dechlorinated water)

d)   sterile cotton

e)   alcohol

f)   burner (either Bunsen or alcohol)

g)   tongs

Procedure

a)   Read the appropriate sections of *Millipore Experiments in Environmental Microbiology*, Millipore Corp., 1970, to become familiar with the equipment and aseptic technique.

b)   Prepare the Swinnex-25 filter holder with a Type GS filter.

c)   Sterilize 6 dilution tubes by boiling them for 3 minutes. Stand these in the dilution tube rack.

d)   Sterilize the Sterifil apparatus by boiling all of the parts for 3 minutes. Prepare it with a Type HAWG filter.

e)   Using a plastic syringe, filter 3–4 ml of dechlorinated water through the Swinnex into 5 of the dilution tubes. When the syringe needs refilling, remove the Swinnex and draw the distilled water into the syringe. Then replace the Swinnex.

f)   Label these 5 tubes No. 2 through No. 6.

g)   In the unlabeled dilution tube, place enough of the topsoil to fill the tube $\frac{1}{3}$ full. Fill it to within 2 cm of the top with water filtered through the Swinnex. Label this tube No. 1.

h) Plug tube No. 1 with sterile cotton and shake it well. Let it stand for a few minutes in the tube rack to settle.

i) Flame-sterilize a dilution loop. Transfer one loopful of water from tube No. 1 to tube No. 2. Swirl the tube to mix it.

j) Flame the loop. Transfer a loopful of water from tube No. 2 to tube No. 3 and swirl to mix. Continue in sequence until all the tubes have been inoculated. Remember to flame the loop before each use.

k) Pour the contents of tube No. 6 into the Sterifil funnel. Filter the contents through the first test filter.

l) Prepare a petri dish and pad with Total Count Medium (yellow).

m) Using flamed forceps, transfer the test filter from the Sterifil to the pad, and close the dish. Label the petri dish No. 6.

n) Install a fresh Type HAWG filter in the Sterifil. Filter the contents of tube No. 5. (The Sterifil does not have to be sterilized between filtrations. You are filtering the sample with the lowest number of bacteria first and working up to the highest number.)

o) After the filtration of tube No. 5, remove the test filter and place it on a different Total Count pad in a new petri dish. Label this dish No. 5.

p) Installing new filters each time, repeat the procedure for tube No. 4 and tube No. 3. Tubes No. 2 and No. 1 are omitted, because they will have far too many colonies to count.

q) Incubate the petri dishes for 48 hours at room temperature.

r) Record the number of colonies on each petri dish.

s) Repeat the procedure using the subsoil sample.

Discussion

How does the presence of organic material appear to affect bacterial populations? Why? Are the types of colonies seen on the topsoil and subsoil plates substantially different? Account for your answer.

Obtain a sample of topsoil and divide it in half. For 2 weeks keep one-half of the sample moist, but let the remainder dry out. Repeat the procedure using these two samples. Account for the results.

5.37 RELATIONSHIPS BETWEEN SOIL MICROORGANISMS IN THE FIELD

When microorganisms are separated from soil in the laboratory, many relationships between them cannot be observed. The only place where this can be effectively done is in the field. The following exercise is a very simple method of observing relative growth rates, competition, and succession in bacterial and fungal populations in soil.

Materials

a) microscope slides
b) cellophane
c) rubber bands

Procedure

a) Place a piece of cellophane around a microscope slide. Secure it with a rubber band at each end. Make enough of these to place two at each site.

b) Place a sheet of cellophane between two other microscope slides. Secure with a rubber band at each end. Make two of these for each site.

c) Insert these slides into the soil at each of several areas, for example, an eroded area, a deciduous forest, a garden, a grassland area.

d) Determine the moisture content (Section 5.3) and the organic content (Section 5.10) at each site.

e) Observe the slides every 5 days for 3 weeks. Record your observations, including the number of each type of colony and its position on the cellophane.

Discussion

Do some types of colonies appear only in areas of high organic content and/or high moisture content? Which colonies always appear near each other? What evidence do you have of competition between the microorganisms? Is there evidence of succession?

5.38 SOIL MICROORGANISMS AS DECOMPOSERS

Decomposition of materials is essential to provide a sufficient supply of nutrients in the soil. Most of the decomposing action is carried out by microorganisms in the soil. Two groups of soil microorganisms which carry out this necessary task are bacteria and fungi. All objects do not decompose at the same rate. The rate depends partially on the soil in which the objects are found and on the substances of which they are made. This exercise allows you to investigate the effect of these factors on decomposition.

Materials

a) sand

b) soil rich in organic material

c) gravel

d) dead plant material (leaves, grass)

e) dead animal material (household meat, dead insects)

f) processed foodstuff (dry cereal)

g) natural fibers (wool or cotton cloth)

h) synthetic fibers (polyester or nylon cloth)

i) styrofoam cup fragments

j) 12 clay flowerpots, diameter 5 cm

k) 6 petri dishes and lids, diameter 4 cm

l) large tray, dimensions approximately $30 \times 20 \times 5$ cm

---

Procedure

a) Place a small amount of gravel in the base of each pot.

b) Fill 6 of the flowerpots with sand up to 2 cm from the top. Be sure to pack the sand in the pot.

c) Similarly, fill the remaining pots with the soil rich in organic material.

d) Place dead plant material on one pot filled with sand and on one filled with rich soil.

e) Cover each of these 2 pots with an upside-down petri dish or lid. Press them lightly into the soil.

f) Repeat steps d) and e) using the other five material samples.

g) Place all 12 pots in the large tray. Add water to the tray to a depth of at least 3 cm. Maintain this depth throughout the observation period.

h) Keep the tray and its contents in a darkened area at room temperature.

i)   Observe the pots every 5 days (for up to 8 weeks) until changes are obvious in most of the pots. Be sure to test the strength of the cloth fibers from time to time. Record your results.

Discussion

Which materials decomposed most quickly? Are there any materials which appear not to have decomposed? How is the soil organic content related to the decomposition rate? Account for this.

What advice do you have for the "litter-bug"?

5.39   THE SPECIAL RELATIONSHIP BETWEEN LEGUMES AND *RHIZOBIUM*

Some species of bacteria live in symbiosis with other organisms. This means that their relationship is mutually beneficial. One such relationship exists between legumes and bacteria of the genus *Rhizobium*. In the following exercise, you can observe these bacteria and determine how they and the legumes benefit one another.

Materials

a)   fresh clover plant (including roots carefully removed from the soil)

b)   glass microscope slides

c)   microscope

d)   methylene blue stain

e)   9 small flowerpots filled with similar soil

f)   autoclave or oven

g)   clover seeds (soaked for 24 hours in water)

Procedure

a)   Wash off the roots of the clover plant. Observe the nodules there.

b)   Squash a nodule between 2 glass slides.

c)   Place a thin layer of this tissue in a drop of water on a slide. Cover it with a cover slip.

d)   Add a drop of methylene blue stain at the edge of the cover slip. Observe the staining of the bacteria, using the high power lens of the microscope. (An oil immersion lens would be an advantage.)

e) Sterilize the soil-filled pots in the autoclave.

f) Add 2 gm of sodium nitrate to 3 of the pots. Label them.

g) Crush 6 nodules from the clover plant. Suspend them in water. Innoculate 3 other pots equally with this suspension. Label these pots.

h) Label the remaining 3 pots as controls.

i) To all 9 pots, add an equal number of soaked clover seeds.

j) Observe over a 2-week period and record your results.

Discussion

Compare the growth of the seedlings in the nitrate-treated soil with that of the seedlings in the control. Account for any observed differences. Similarly, compare the growth of the seedlings in the soil inoculated with *Rhizobium* with that of the seedlings in the control. Account for any observed differences. Finally, compare the growth of the seedlings in the nitrate-treated soil with that of the seedlings in the soil inoculated with *Rhizobium*. Account for any observed differences.

Explain carefully what is meant by symbiosis, using this exercise to illustrate your explanation.

Farmers commonly alternate a crop of clover with a crop of wheat in the same field. Why do they do this?

More detailed studies of this topic are available in several books. These studies involve the preparation of special media and the use of bacterial cultures. Two such books are *Microbes in Action—A Laboratory Manual of Microbiology*, H. W. Seeley, Jr. and Paul J. Vandemark (W. H. Freeman, San Francisco, 1962), pp. 160-164, and *Life in the Soil*, David Pramer (D. C. Heath, Boston, 1968), pp. 54-56.

5.40 FORMATION OF HUMUS IN A COMPOST HEAP

Compost heaps have been used by gardeners for many years as a private source of humus. Easily made and maintained, these piles rely on decomposition processes. By this method, nutrients from dead plant material are recycled in the garden. Organic materials are acted upon by soil organisms such as earthworms, nematodes, and bacteria, until the plant material is reduced to humus. Various physical and chemical factors change during this process. It is interesting to measure these factors and to relate the measurements to the biological processes that are occurring.

Materials

a) various types of dead plant material, for example, leaves, grass clippings, straw, table scraps (but no animal fats), sawdust, hedge clippings, weeds

b) pH testing materials (see Section 5.12)

c) soil thermometer

d) materials for measuring soil organic content (see Section 5.10)

e) 12-12-12 fertilizer

Procedure

a) Construct a frame in which to build your compost heap by digging a pit, or by setting up a rectangular frame of boards, snowfence, or wire.

b) Place 20–30 cm of plant material in the frame.

c) Cover this with a layer of 12-12-12 fertilizer, using approximately 1 liter of fertilizer for every 3 cubic meters of plant material.

d) Moisten this thoroughly. Then cover it with 2 or 3 cm of soil.

e) Make a depression in the top to catch rainwater.

f) Turn the pile once each month and water it thoroughly.

g) Before turning the pile record the temperature, pH, and organic content at the center of the pile. Record your results.

Discussion

Describe and account for changes in appearance. Account for any changes in temperature which you observe in the heap. Did the pH change in the manner you expected? Explain. Account for the results you obtained when you determined the organic content at the center of the pile.

Complete conversion to humus in a compost heap takes about a year. Devise a method by which your compost heap can be kept in continuous production, with new material being added and humus being removed at any time. Recycle the humus by placing it on the school lawn and gardens.

Explain the meanings of the terms *litter*, *fermentation*, and *humus*, as illustrated by this study.

5.41 PHYSICAL AND CHEMICAL ASPECTS OF SOIL PROFILES

This is a major field study that requires performing many of the preceding exercises at several sites. If you have never studied a profile before, choose one site where the soil is well drained. (The horizons will usually be easier to distinguish there.) Choose further sites at a variety of locations. You could compare sites in a deciduous and a coniferous forest, or a deciduous forest and a grassland, or various sites along a slope to see the effects of drainage. An aerial photograph of the area which you intend to visit is very helpful in choosing the study sites. Topographical maps and soil maps are also helpful in this respect. These materials are generally available from government agencies. Predict what types of profiles you will find at each site.

Materials

a) shovels or soil samplers

b) measuring tapes

c) *p*H testing materials

d) dilute hydrochloric acid

e) eye dropper

f) LaMotte Soil Testing kit

g) materials for determining moisture content (see Section 5.3)

h) materials for determining water-holding capacity (see Section 5.4)

i) materials for determining organic content (see Section 5.10)

j) clinometer (for determining slope)

Procedure

a) Prepare a table with the following column headings: Site, Horizon, Depth, Color, Texture, Structure, *p*H, Presence of calcium carbonate, Mineral content, Moisture content, Water-holding capacity, Organic content. Make enough copies of this table for recording observations at each profile.

b)   Observe and record the topography (slope) of each site and the types of vegetation present. Record the date and the weather conditions during the study and during the 24 hours that preceded the study.

c)   Your soil profiles can be exposed in one of two ways— either dig a pit, or use a soil sampler. (Both methods are outlined in Section 5.1.)

d)   Record the depth to which the majority of roots descend in each profile. Note the depth at which good aeration appears to cease.

e)   At each profile, identify each horizon and mark its boundaries with sticks, nails, or some other markers.

f)   Perform steps g) through o) at each site. Enter the results in the prepared table.

g)   Measure the thickness of each horizon.

h)   Observe the color of each horizon when the soil is wet, and when it is dry (if possible).

i)   Observe the texture and structure of the soil in each horizon.

j)   Find the pH of each horizon (Section 5.12).

k)   Determine whether calcium carbonate is present. Using an eye dropper, place a few drops of dilute hydrochloric acid on each horizon. The presence of calcium carbonate is shown by bubbling. This is the result of a chemical reaction which produces carbon dioxide. Calcium carbonate easily leaches out of soil. Thus the horizon containing the most calcium carbonate is at the average depth to which percolating water descends. This, then, is an indication of how much leaching from the higher horizons is occurring.

l)   Determine the mineral content (nitrogen, phosphorus, potassium, and calcium) of each horizon (Section 5.13).

m)   Determine the moisture content of each horizon (Section 5.3).

n)   Determine the water-holding capacity of each horizon (Section 5.4).

o)   Determine, by ignition, the percent organic content of each horizon (Section 5.10).

Discussion

Back in the classroom, try to analyze the information you have obtained. How much leaching is occurring at each site? Has this affected the vegetation? Which of the sites would erode most easily? Support your answer. What is the importance of depth of root growth and of depth of aeration in soil? What relationships exist between the nature of each profile and the vegetation in the area?

5.42 BIOLOGICAL ASPECTS OF SOIL PROFILES

This study is best carried out in conjunction with Section 5.41, since both are long studies. One class could carry out Section 5.41 and another class, Section 5.42. Results can be shared for discussion purposes. If this study is carried out independently of Section 5.41, the sites should be chosen in such a manner that profiles differ significantly from one another.

Again, soil and topographical maps should be studied carefully before this study. Predict what types of vegetation can thrive at each site. What relative sizes of soil animal populations do you expect to find at each site and in each horizon?

Materials

a)   shovels or soil samplers

b)   measuring tapes

c)   materials for separating arthropods (see Sections 5.27 and 5.28)

d)   materials for separating nematodes (see Section 5.34)

e)   materials for separating protozoa (see Section 5.35)

f)   materials for separating bacteria (see Section 5.36)

g)   surface traps (see Section 5.23)

h)   microscope slides (see Section 5.37)

i)   containers

j)   labels

k)   notebooks and pencils

Procedure

a)   Make up a data sheet with the following column headings: Site, Horizon, Depth, Arthropod number, Nematode number, Protozoan presence, Bacterial variety and number. Make enough copies of this sheet for recording data at each site.

b) Record the date, the slope of the land, and the visible vegetation at each site. Note the weather conditions that prevailed during the study and for the 24 hours preceding the study.

c) If not previously done in Section 5.41, the soil profiles can be exposed in one of two ways—either dig a pit or use a soil sampler (Section 5.1).

d) Record the depth to which good aeration appears to be present. Note how deep the majority of roots descend at each profile.

e) At each site, identify each horizon and mark its boundaries with sticks, nails, or some other markers.

f) Measure and record the thickness of each horizon.

g) Set up the surface traps and the buried microscope slides in the vicinity of each exposed profile. Return at the prescribed time and record the species (if possible) and the numbers in each type of trap.

h) Remove large enough samples of each horizon at each site to carry out steps i) through l) in the laboratory. Enter the results in the prepared data sheets. Be sure to note the size of the soil sample from which the organisms were separated.

i) Using a Tullgren funnel, separate the arthropods from each horizon sample. Record the number of each species, if possible.

j) Using a Baermann funnel, separate the nematodes from the samples. Record the number in each horizon at each site.

k) Determine whether protozoa are present or absent in each horizon.

l) Using the Millipore method, separate out the bacteria. Note their variety and numbers.

Discussion

Now that you have accumulated large amounts of information, it makes sense to try to find relationships in it. However, be careful not to draw conclusions when you do not have sufficient information.

Describe and account for the diversity of species that exists in each horizon of each site. Why is the variety of species present as important or even more important than the total number of organisms?

If you completed Section 5.41, determine how the physical and chemical conditions of the soil relate to the populations in each horizon of each site. Do they affect the surface vegetation noticeably? Is there any evidence of the impact of man on any of your sites?

5.43 THE STATE OF HEALTH OF YOUR SCHOOL YARD

Since this study will take place on the school property, it must involve procedures which are not likely to mark the property noticeably.

Materials

a) soil samplers

b) containers and labels

c) materials for measuring percolation rate (see Section 5.6)

d) materials for measuring water content (see Section 5.3)

e) materials for measuring water-holding capacity (see Section 5.4)

f) pH testing materials (see Section 5.12)

g) LaMotte Soil Testing kit

h) materials for measuring organic content (see Section 5.10)

i) materials for measuring capillarity (see Section 5.7)

j) materials for measuring pore space (see Section 5.8)

k) notebooks

Procedure

a) From the maintenance staff, learn the age of the school excavation, whether the school was recently landscaped, what type of fertilizer is used on the soil, and what type of grass has been planted on it.

b) Draw up a table with the following column headings: Site, Rate of infiltration, Moisture content, Water-holding capacity, pH, Mineral content, Organic content, Capillarity, Pore space. Make enough copies for use at all the sites.

c) Map the school property. Then choose at least five sites in such a way that they are evenly distributed throughout the property.

d) Record the date, weather conditions, and general appearance of the vegetation in the area of each sampling site.

e) Determine the rate of infiltration at each site. Record the results.

f) No pits should be dug for this study. Sampling should be done in the top 10–15 cm of soil with a soil sampler. (Replace all your divets, please!) Collect your samples and perform as many of the following tests as possible: water content; water-holding capacity; pH; mineral content (nitrogen, phosphorus, potassium, and calcium); organic content; capillarity; and pore space.

Discussion

What fertilizer should be used on the property? Should the lawn-watering practices be changed? What plants might be better suited to this soil? Would you expect many earthworms to be present, considering the pH, water content, and organic content of the soil? Is the soil too compact?

Prepare recommendations for maintenance of the school yard and submit them to the proper authorities.

You may wish to carry out a similar study on your home lawn or in a field on a farm.

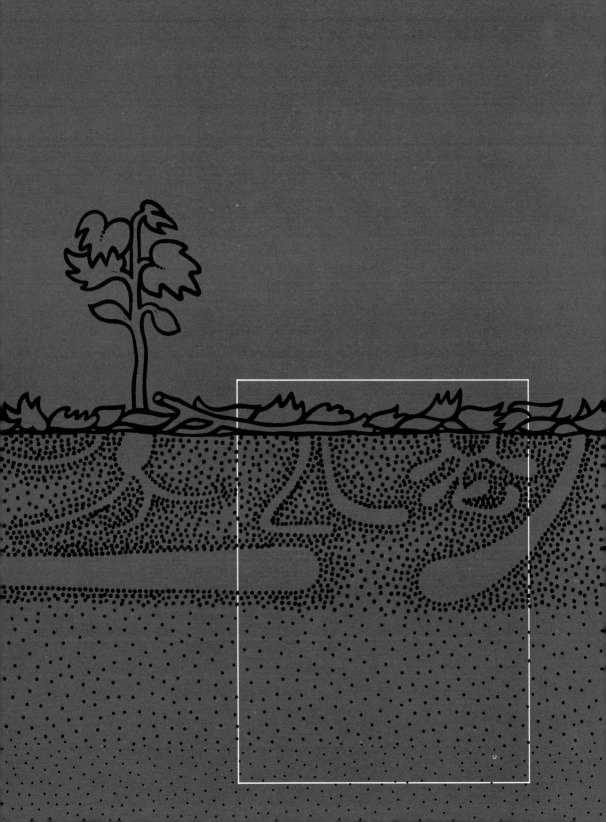

Research Topics

6

The topics in this Unit range from very practical problems like soil erosion, to hypothetical problems like farming the moon. None of these topics has been discussed to any great extent elsewhere in this book. You are to examine the topics and select one that interests you. Then research it thoroughly, either by yourself or with a small group of others who are interested in the same topic.

A list of suggestions *For Thought and Research* has been included with each topic. This list was included to help you to get started with your project. It is not a complete list of things you should do. As you research the topic, you will think of many further things to add to this list. A few references have been included with each topic. You should also consult encyclopaedias and other books in your school and municipal libraries. Government agencies such as your Department of Agriculture will have information on some of these topics.

6.1   SOIL EROSION

The soil over ever-increasing areas of North America is being rendered useless for agricultural purposes. A great deal of the fertile topsoil has vanished, leaving behind a cement-like surface of packed soil and pebbles. This deterioration of our soil has been caused, in part, by erosion. The forces most responsible for this erosion are wind, water, and ice.

Years ago, the leaves of the forests and the grasses of the prairies broke the force of the pounding rains, allowing the water to drop for hours without damage. Humus formed by decaying leaves and grass soaked up much of even the heaviest rains. But the trees have been removed and the grasses plowed under to make way for more agricultural land. Now, in many areas, the bare ground, sparsely covered with vegetation, takes the full pounding of the rain.

Whenever rain falls on bare earth, denuded by poor agricultural practices, overgrazing, or fire, it cannot soak in, but runs off, carrying a large amount of the soil with it. This is the beginning of water erosion. This situation can be further aggravated until even light rains cause rushing gullies of silt-filled water. As more soil is washed from a region, the remaining soil becomes less able to absorb moisture. Thus the erosion becomes increasingly more serious with each succeeding rainfall. Exposed hills on steep slopes are extremely susceptible to this form of erosion (Fig. 6-1). Also, when soil in this condition dries out, it can often be blown into sand and dust storms even by normal winds.

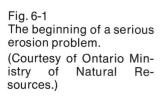
Fig. 6-1
The beginning of a serious erosion problem.
(Courtesy of Ontario Ministry of Natural Resources.)

The effects of this erosion are numerous. They include loss of soil and soil fertility, ruined and unproductive fields, and loss of water as runoff. This can lead to crop failures; decreased land values; silt build-up in lakes, streams, and reservoirs; abandoned farms; and destruction of wildlife.

The fact that water easily runs off eroded soil causes a multitude of further problems. During periods of heavy rainfall, the rivers in the area become flooded with water that is flowing over the soil rather than being slowly absorbed by it. Dangerous flash-floods often occur. When little or no rain falls, droughts occur since the soil retains little moisture. Streams and rivers whose drainage basins experience alternate periods of high and low water become virtually useless for navigation, town water supplies, hydroelectric power, and irrigation. The costly measures that must be taken by governments to counteract the effects of erosion can be avoided only if the earth's surface can absorb and hold most of the rainfall in reserve beneath the surface.

For Thought and Research

The following are some aspects of soil erosion that you should investigate.

1 The magnitude of the problem in North America. How much soil do we lose per year?

2 The factors that determine the amount and rate of soil runoff.

3 The methods commonly used to prevent or reduce soil erosion.

4 The advantages and disadvantages of strip cropping.

5 The effects of fire on soil mineral and organic matter, and how this ultimately increases the danger of erosion.

6 Trace erosion by water through its normal sequence (from sheet erosion to gullying).

7 It has been stated that to control erosion permanently it is necessary to treat the cause rather than the consequences. Discuss this statement.

8 Locate the areas of worst soil erosion in North America. Give the reasons for erosion in these areas.

9 Once the soil disappears from wind or water erosion, the climate tends to become progressively more extreme. Why is this so?

Recommended Readings

1 *Basic Ecology* by Ralph and Mildred Buchsbaum, Boxwood Press, 1957. The first 16 pages of this book provide an excellent account of soil erosion.

2 *Conserving Soil* by M. D. Butler, Van Nostrand, 1955. A good general discussion of erosion and methods of prevention.

3 *Profitable Soil Management* by L. L. Knuti, M. Korpi, and J. C. Hide, Prentice-Hall, 1970. See Chapter 17 for a discussion of erosion problems in the United States.

4 *Soil Conservation* by S. G. Archer, University of Oklahoma Press, 1960. Contains very readable sections on soil erosion and methods of treatment.
5 *The Rape of the Earth* by G. V. Jacks and R. O. Whyte, Faber and Faber, 1947. This book presents a large and complete study of soil erosion and a multitude of related factors. Also contains an erosion map of the United States.
6 *Vegetation and Watershed Management* by E. A. Colman, Ronald Press, 1953. Explains how the management of vegetation helps to control the water in streams and ground-water basins.

6.2 SOIL POLLUTION

Soil pollution, while not posing an immediate threat to humans, does worry many biologists because of the possible long-term ecological effects. Since the soil produces a great deal of our food, it is necessary to have it as productive as possible. Rich, productive soil is full of microorganisms which decompose the organic matter, giving the soil its richness. If soil pollution decreases or eliminates these microorganisms, a poorer soil and hence poorer crops will be produced.

Soil pollution is caused by many substances. The most common are pesticides, lead emitted from automobiles, and salt used in Canada and many of the northern states to melt the ice on streets and highways.

Pesticides are chemicals used to eliminate undesirable plants, animals, or microorganisms. Foremost among pesticides are insecticides (insect killers), herbicides (plant killers), and fungicides (fungi killers). If an ideal pesticide could be produced, it would kill only one particular species. Unfortunately no such pesticide exists. In fact, most are highly unselective, and many are poisonous to humans.

As the human population of the earth increases, more and more food is required. Since pesticides provide increased yields of higher quality food, their use is also increasing. Almost everyone in North America uses pesticides. Two common uses are to kill weeds on lawns and flies in homes. Pesticides are applied as dusts to foliage, as aerial sprays, or directly to the soil. When pesticides are released near the soil surface their movement is not easily followed. Some are carried away with the soil water runoff. Others remain in the soil.

Some of the stored pesticides break down readily in the soil into harmless substances, while others tend to persist in the environment. Generally speaking, the effects of these accumulations of pesticides on populations of beneficial soil organisms and on the chemical properties of soils are not well known. Since they are applied to a complex ecosystem, it is almost certain that they kill or affect organisms other than the target species (that organism for which the pesticide was applied). This may have far-reaching ecological effects. If a population of organisms is not able to re-establish itself, either through surviving organisms or through organisms from nearby untreated areas, it may be entirely replaced by other, less useful forms. Other organisms may then increase in numbers because their principal predators are killed. In this way, the whole ecology of the soil in that area is altered.

Because many pesticides persist as residues in the soil for many years, they can be ingested by animals, stored in body tissues, and passed from one animal to another along the food chain. Thus animals at the end of food chains frequently have pesticide levels many times greater than the level resulting from an individual application of the pesticide. These residues can be harmful to the animal, often resulting in death. Since man is at the end of many food chains, this could be an indication of the long-range effects of pesticides on people.

A good example of this accumulation of pesticide residues through food chains has occurred in California. Many fish-eating birds such as white pelicans, common egrets, and great blue herons have died as a result of the accumulation of pesticides which were passed to them along a food chain involving invertebrates and fish. The effects of the movement of pesticides into streams and lakes was also noted recently in New York state. Here, lake trout production was completely halted in some lakes. Apparently healthy female trout had enough pesticide in their eggs to cause the death of all of the hatching fish. The elimination of many species of wildlife by pesticide poisoning is indeed a saddening and immediate possibility.

Soil pollution from salt and lead does not appear to be as serious as that from pesticides. Salt does, however, act as a herbicide and may have other effects on the soil. It appears that lead is being concentrated in vegetation along our roads. This problem should be carefully researched, since lead is a highly toxic substance.

For Thought and Research

The following are some aspects of soil pollution that you should investigate.

1 Prepare a complete list of the harmful aspects of chemical pest control. Prepare a similar list of the beneficial aspects. Now prepare either an argument that defends the use of pesticides or one that recommends discontinuing their use. Alternatively, you may want to propose a compromise.

2   Investigate sources of pesticide pollution in your vicinity. Find out what types of pesticides are used, when they are used, and how often they are applied during the year. Also find out the concentration of the pesticide used and how the target species is affected.

3   Research the laws in your area governing pesticide use. Which pesticides have been recently banned and why?

4   List 10 commonly used pesticides, and give for each the rate of application and the target species for which it is used. Also describe for each the effects it has on humans and list recommended safety precautions. A trip to a hardware store or garden center would be beneficial here.

5   Find out what substitutes exist for pesticides.

6   How does a species develop resistance to a particular pesticide? What danger does this present in humans?

7   What are *secondary poisoning* and *delayed expression*?

8   List the factors that speed up the loss of pesticide residues in the soil. What happens to these residues?

9   India uses vast amounts of DDT. Why? Suppose India were to ban the use of DDT as have many western countries. What would be the possible consequence of this ban?

10   Countless tons of herbicides have been used as defoliants in Vietnam. What are defoliants? How do they work? Do they accumulate in the soil? What are the long-term effects on the ecology of an area?

11   Make a list of the five most persistent pesticides that are still sold in your province or state. Are any of these on a restricted list?

12   DDT is one of the most persistent and dangerous of all pesticides. As a result, its use is gradually being banned throughout North America. For example, in Ontario, tobacco farmers are now virtually the only people who can obtain a permit to use DDT. Why do you suppose they are granted this permit? If you do not live in Ontario, find out if a comparable situation exists in your state or province. Your Department of Agriculture will have the information.

Recommended Readings

1   *Defoliation* by Thomas Whiteside, Ballantine Books, 1970. An interesting book on the use and effects of the herbicide 2,4,5-T in Vietnam and in the United States.

2   *Ecology and Field Biology* by R. L. Smith, Harper & Row, 1966. Read pp. 366 and 367 for the results of some recent studies on pesticide pollution.

3   *Pesticides and the Living Landscape* by R. L. Rudd, University of Wisconsin Press, 1964. An excellent and comprehensive book on this topic. It explains the types of pesticides and gives reasons for and against their use.

4   *Pesticides and Pollution* by S. C. Bloom and S. E. Degler, Bureau of National Affairs Environmental Management Series, 1969. Presents an account of pesticide pollution throughout North America.

5   *Pesticides and Their Effects on Soils and Water*, A.S.A. Special Publication Number 8, Soil Science Society of America, 1966. Good discussion on the effects of pesticides on the ecosystem. Also considers other means of pest control.

6   *Persistent Pesticides in the Environment* by C. A. Edwards, Chemical Rubber Company, 1970. Considers removal of pesticide residues as well as pesticide substitutes.

7 *Silent Spring* by Rachel Carson, Houghton Mifflin, 1962. A classic work on pesticides and their potential effects on man and wildlife.

8 *Since Silent Spring* by Frank Graham, Jr., Houghton Mifflin, 1970. Provides a guide for safer home and garden use of pesticides as well as an appendix of those which should be avoided.

9 *Some Safety Aspects of Pesticides*, Proceedings of a Conference at the British Museum (Natural History), London, 1967.

10 *The Pesticide Problem* by J. C. Headly and J. N. Lewis, Johns Hopkins Press, 1967. Discusses benefits and disadvantages of pesticide use.

11 *The Environmental Handbook* by Garrett De Bell (editor), Ballantine Books, 1970. See the chapter "Pesticides Since Silent Spring" by S. H. Wodka for an excellent review of DDT and the herbicides 2,4,5-T and 2,4-D (with information on their effects on humans).

12 *Cleaning Our Environment. The Chemical Basis for Action*, American Chemical Society, 1969. The section on pesticides in the environment describes numerous pesticides and discusses their effects and persistence. Factors in the soil that affect persistence are also discussed.

6.3   SOCIAL BEHAVIOR IN ANT COLONIES

An ant colony is an ecological phenomenon which succeeds because of unique behavior patterns. Division of labor is the mainstay of an ant colony. Ants may specialize in warfare, communication, slave ownership, gardening, dairying, or nest construction, just to name a few.

Scientists have been fascinated by the complex behavior of ants as they instinctively perform their designated roles. This research topic is designed to let you get a glimpse of the intriguing life of ants and to see how their behavior has enabled ants to live in practically every possible environment, except those where the soil is permanently frozen or flooded.

For Thought and Research

The following are aspects of ant behavior that you should find interesting to research.

1 The anatomy of ants.

2 Castes and classes among ants.

3 The formation of new colonies.

4 The warfare of legionary or driver ants such as South America's *Eciton hamatum*.

5 The "slave raiders."

6 The reasons why *Iridomyrmex humilis*, the Argentine ant, is currently spreading over the world, conquering one ant species after another.

7 Granaries and fungus gardens of leaf-cutter ants.

8 Aphid tenders and honey ants.

9 The ant-eater's adaptations as a natural enemy.

10 Solomon, in his biblical Proverbs, states, "Go to the ant, thou sluggard: consider her ways and be wise." Is it possible to compare the actions of ants and men? Do humans have the civic virtues of ants? Is any one of man's forms of government (imperialistic, democratic, or socialistic) comparable to that of an ant colony?

Should humans act or do we, in fact, already act like ants, only on a larger scale?

Recommended Readings

1 *Our Insect Friends and Foes* by W. A. Depuy, Dover, 1968, pp. 122-137. An interesting and readable section on ants.

2 *The Ants* by Wilhelm Goetsch, University of Michigan Press, 1957. This book is excellent, all-inclusive, and perfectly clear. It is probably the best source of information on this topic.

3 *All About Ants* by P. P. and M. W. Larson, World, 1965. A thorough account of ants and ant behavior.

4 *Ants from Close Up* by L. H. Newman, Thomas Y. Crowell, 1967. Contains excellent illustrations on the subject.

5 *Insects* by R. E. Hutchins, Prentice-Hall, 1972. Contains an interesting and informative description of the social behavior of ants.

6.4   SOIL ON THE MOON AND PLANETS

The great changes and advances in the twentieth century have led many people to believe that technology can solve any problem. However, others are beginning to voice a different view—that technology could eventually lead to man's downfall through the destruction of his own environment. Optimists on the side of technology have stated that, in time, man will be able to travel to other planets to escape earthly problems like pollution and over-population.

Without getting too far into the argument, let us assume that technology will make it possible for man to migrate to another planet. What practical problems would be involved in setting up a permanent colony there? Could we develop agriculture and produce the foods we are accustomed to on earth? Would the soil be the same valuable resource that it is here? Would the same recycling take place on the moon that takes place on earth? What things would have to be done to establish a viable soil capable of maintaining the same complex balance seen on earth?

For Thought and Research

Begin this research topic by investigating these areas.

1 What has been learned so far about the soil on the moon? Find out what you can about the following characteristics of moon "soil": its mineral composition, texture, water content, temperature extremes, and organic content.

2 What would have to be done to create a productive soil environment on the moon? Would the nutrient cycles operate on the moon?

3 Find out what sort of conditions exist on Mars and what the soil of this planet is probably like.

4 Repeat question 3 for Venus.

5 Research the surface characteristics of the other planets in the solar system. Rank them according to their potential use in starting a permanent human colony.

6 Other than the problems associated with soil, what other barriers are there to extensive colonization of the moon and planets by humans?

Recommended Readings

1 *Life on the Planets*, by R. Tocquet, Grove Press, 1962. This small paperback starts with basic conditions necessary for life and discusses, in turn, the potential for life on the moon, Mars, Venus, and all other planets. The best book for the layman.

2 *Planets* by C. Sagan and J. N. Leonard, Life Science Library, Time, Inc., 1966. Chapters 5 and 6 are excellent on Venus and Mars.

3 *Earth, Moon, and Planets* by F. L. Whipple, Harvard University Press, 1968. This book is rather technical, but contains a great deal of useful information on the topic.

4 *National Geographic Magazine*, May, 1963; February, 1969; December, 1969; July, 1971; February, 1972.

6.5 RECYCLING OF ELEMENTS

> . . . the hands of the sisters Death and Night
>   incessantly softly wash again, and
>   ever again, this soil'd world. . .
>
> Walt Whitman

Death is a basic fact of life, an inevitable consequence of living. The atoms and molecules making up any organism must eventually be returned to the lifeless state.

Just think, every square centimeter of the earth's surface has probably at some point in time been the death bed of some organism. But the earth's surface is not knee-deep in the dead remains of plants and animals. The earth today is covered with active, vibrant life. The bodies of dead organisms decay and are reduced to the gases, liquids, and minerals of which they were made. The atoms of lifeless organic matter are quickly incorporated into new life or are released as simple gases, such as carbon dioxide, or liquids, such as water. Eventually these gases and liquids may be incorporated into new living tissue.

The purpose of this research exercise is to emphasize the process by which matter is recycled after death to be returned to the environment and possibly reused by other life. Your assignment is to trace the imaginary history of a single atom of carbon through thousands or even millions of years here on earth. At this very moment, the carbon atom you wish to consider may be locked up in the wood fibers of your desk or in the protein molecules of one of your eyelashes. In the past, this same atom may have dwelt for limited periods of time in the bodies of many animals and plants, perhaps even in some huge prehistoric dinosaur. The possible histories of any given carbon atom are as numerous as your imagination and understanding of the world permit.

For Thought and Research

Investigate questions 1 through 6 before you try to trace the path of a carbon atom through history.

1 Why is carbon so important to living organisms?

2 Trace the path taken by carbon dioxide as it is removed from the air and converted into organic substances in a plant. Explain how the gas enters the plant, how it gets inside the cells, and what happens to it in the cells.

3 Outline two or three hypothetical food chains through which the atom might pass. At least one of these should be in the soil ecosystem.

4 Using earth history texts, determine when life first appeared on earth. Summarize the evolutionary time scale.

5 Find out when large amounts of carbon were locked up in the underground reserves of coal and oil. What conditions prevented recycling of dead organic matter in ancient soils, thus giving rise to huge coal and oil deposits?

6 How can the age of a fossil be determined using carbon 14?

7 Pick a carbon atom, for example, one in your eyelash. Where was it before it entered your eyelash? And before that? Continue this process as long as you can. Pay particular attention to the role of the soil in the determination of the life history of your carbon atom.

If you want to make this a truly worthwhile and interesting research exercise, trace your historical account right back to when the earth was first formed. Describe each environment that the atom enters in as much detail as possible, especially if it enters the soil community, which it would most likely do on a number of occasions. If you are good at telling tall tales, here is your chance to let your imagination run wild.

Make sure your teacher sets some sort of word limit to your essay. Otherwise it could go on forever.

Recommended Readings

1 *Historical Geology* by C. O. Dunbar and K. Waagé, John Wiley & Sons, 1969. Read Chapter 6, "The Constant Change of Living Things." The rest of the book is mainly geology-oriented, but contains information on extinct life forms.

2 *Man and the Natural World* by C. J. Goin and O. B. Goin, Macmillan, 1970. An extremely useful text, especially Chapters 23-27, which deal with evolution, geologic time, and the history of life forms.

3 *The History of Life* by A. L. McAlester, Prentice-Hall, 1968. An excellent, in-depth book.

4 *You and the Universe* by N. J. Berrill, Dodd, Mead, 1960. Read the chapter entitled "Life and Death" and then you'll probably want to read the whole book.

Case Studies

7

The case studies in this Unit consist of actual scientific data collected during the course of experiments. They are included to give you a chance to find out if you can apply your knowledge of soil ecology. By now you should understand clearly the ecosystem concept. You should be particularly aware of the close interdependence that exists between biotic and abiotic factors in soil ecosystems. If the number of any one animal species is altered in a sample of soil, the numbers of all other species will likely change. If a physical factor like the moisture content is altered, many other physical factors like soil temperature and oxygen content may change. This, in turn, will affect the living organisms in the soil.

     The chain of events that may be triggered by one act is unending and very complex. But it is certain to occur. You have undoubtedly discovered this in your field and laboratory studies. Try these case studies to see if you can apply your knowledge of soil science and the ecosystem concept to data collected by others.

7.1   DECOMPOSITION OF LEAF LITTER (I)

The decomposition of organic matter is a very important process in the soil ecosystem. It supplies energy to soil heterotrophs and forms a necessary link in the recycling of nutrients.

Decomposition takes a long time. It would not be feasible to station an observer beside a fallen leaf and tell him to record all the changes in that leaf over the next 12 months! In the first place, he could only see what happened while the leaf was above ground. In addition, he would be unable to see the microscopic changes. Besides, who would want such a job? Here is a more practical method for studying leaf decomposition.

Leaves from the red oak, *Quercus rubra*, were collected and placed in bags made of nylon mesh, 20 leaves to a bag. Three bags were used, each having a different mesh size.

Each bag was tied shut, labeled, weighed, and buried at a depth of 10 cm in a flower garden on June 1. At one month intervals, until frost came in November, the bags were dug up and weighed, and then returned to the ground. In April, when the ground had thawed, the routine was resumed. The results appear in Table 3.

TABLE 3

Month	Mesh size		
	10 mm	1 mm	0.005 mm
June	37.9 gm	39.3 gm	41.2 gm
July	22.3 gm	35.1 gm	40.4 gm
August	17.5 gm	32.4 gm	40.0 gm
September	9.6 gm	27.7 gm	38.3 gm
October	5.9 gm	23.2 gm	36.7 gm
November	4.9 gm	21.9 gm	36.2 gm
April	3.3 gm	21.0 gm	35.0 gm
May	2.4 gm	18.4 gm	34.2 gm
June	2.0 gm	16.0 gm	32.9 gm

1 Using the data given, plot graphs to show the change in weight of the buried oak leaves for each of the three bags. Plot time on the *x*-axis and weight of leaves on the *y*-axis. Put all three lines on the one set of axes.

2 Why does the weight of the leaves in all three bags change very little between November and April?

3 (a) What groups of soil organisms can go through a 10 mm nylon mesh? A 1 mm nylon mesh? A 0.005 mm nylon mesh?

(b) What was the experimenter trying to prove by conducting the experiment with bags of three different mesh sizes?

4 Using the graph and your answer to 3(a), formulate a theory about the relative importance of various groups of soil organisms in the process of leaf decomposition.

5 More than three-quarters of the energy flow in an average soil is through microorganisms. (This means that experiments have shown that microorganisms are responsible for most of the decomposition of organic matter.) How do you account for your graph in the light of this statement? Does this explanation differ from the one you proposed in your answer to question 4?

6 Examine carefully the procedure described in this study. What are some possible sources of error?

7 Will the weight of the leaves in the 10 mm mesh bag ever reach 0 gm? Explain your answer.

7.2   DECOMPOSITION OF LEAF LITTER (II)

A researcher, interested in the disappearance of fallen leaves in a deciduous forest, carried out a field experiment that lasted nearly a year. He collected all the leaves from 100 plots scattered throughout the forest. Each plot was 1 meter square. He did this

TABLE 4   TOTAL DRY WEIGHT OF ALL THE LEAVES
COLLECTED FOR EACH SPECIES (gm)

Collection date	Ash	Beech	Elm	Hazel	Oak	Willow
November	4,270 (100)	3,220 (100)	3,481 (100)	1,723 (100)	5,317 (100)	3,430 (100)
May	2,431 (57)	3,190 (91)	1,739 (−)	501 (−)	4,401 (83)	1,201 (35)
August	1,376 (32)	2,285 (71)	35 (−)	62 (−)	1,759 (33)	4 (1)

in November. In May he returned and collected leaves from 100 other plots, each plot just a few meters away from one of the original plots. He did the same thing again in August. After sorting, drying, and weighing all of the leaves, he summarized his findings in Table 4. The numerals in parentheses indicate the percentage of the leaves remaining, using the November values as 100%.

Questions

1 Complete Table 4 by calculating the percentage of elm and hazel leaves remaining in May and August. Record these values in your notebook.

2 Construct a graph as shown in Figure 7-1. Plot the data for November, May, and August for the ash leaves. Join the plotted points with straight lines.

Fig. 7-1
Decomposition of leaf litter.

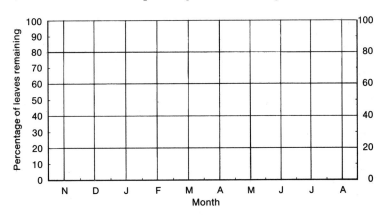

3 Using different colors, plot and draw the lines for each of the other species. Keep track of which lines pertain to which species.

4 Which leaves disappear most slowly?

5 Which leaves disappear most quickly?

6 Why was November a good month to start the experiment?

7 What agents are probably responsible for the disappearance of the leaves?

8 Why did fewer leaves disappear over the first 6 months than in the last 3 months of the experiment?

9 Oak trees have a tendency to hold on to dead leaves well into the winter, sometimes not dropping them until the spring. Could this lead to misinterpretation of the data?

10 Why do some leaves disappear faster than others?

11 What might eventually happen if this forest evolved and became exclusively made up of beech trees?

7.3   SOIL CENSUS OF ARTHROPODS

A soil ecologist was interested in obtaining a census of the arthropods in a permanent pasture grazed by cattle. To do this,

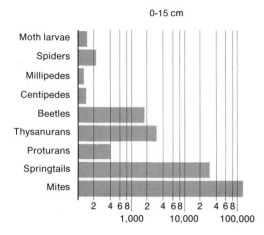

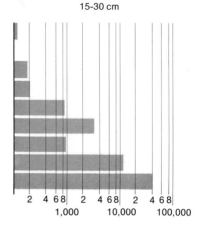

Fig. 7-2
The number of soil arthropods per square meter in permanent grassland soil at two different levels.

study plots 1 square meter in size were established. Soil samples were then taken from the top 15 cm and from the 15-30 cm layer. A Tullgren funnel, among other things, was used to get an accurate count of the numbers of organisms from various arthropod groups that were present at each level. The data are summarized in Figure 7-2.

Questions

1 Make a table consisting of three columns, showing those arthropods which decreased, those that showed no change, and those that increased in numbers from the top layer to the lower layer.
2 In the 0-15 cm layer, the population of springtails is how many times greater than the centipede population? (Note that the scale is logarithmic.)
3 Why do most of the populations decrease in numbers from the top layer to the lower layer?
4 Where would most of the spiders be located within the 0-15 cm layer? Why?
5 Even though more food seems to be available in the top 15 cm, centipedes are more abundant at greater depths. What might explain this distribution?
6 If the collections had been made at night, what changes might have occurred in the results? Why?
7 If the soil ecologist had taken a sample between 30 and 45 cm, would you predict more or less of the following arthropod types: centipedes, beetles, and springtails? Support each prediction with at least one reason.
8 Why might the presence of cattle cause local high densities of certain arthropod populations within the pasture field?

7.4 EARTHWORM ACTIVITY IN DIFFERENT HABITATS

One of the important functions performed by earthworms is the turnover of soil which occurs as a result of their burrowing. Soil that is swallowed is often brought to the surface and piled up as "castings" outside the worm's burrow. The amount of this excrement on the surface gives a good indication of the amount of earthworm activity in the soil.

An ecologist who was studying earthworms and their importance to the soil set up an experiment as follows: One square meter plots were established in eight different types of environments ranging from a golf course to a garden. Worm casts (excrement) were collected each month starting in January and ending in December. The excrement was dried and weighed. The results are tabulated in Table 5.

TABLE 5 EARTHWORM EXCREMENT (gm per m²)

Month	Golf course (always wet)	Woodland meadow (liquid manure each year)	Permanent meadow (liquid manure each year)	Orchard	Permanent meadow (wet)	Spruce forest	Mixed forest	Garden soil
Jan.	—	—	—	—	—	—	—	—
Feb.	—	—	—	—	—	—	—	—
Mar.	120	205	130	180	70	20	39	6
April	980	1,360	475	540	571	160	555	11
May	745	1,402	987	378	685	470	464	—
June	1,210	985	910	382	186	184	314	—
July	688	270	245	300	98	73	115	—
Aug.	711	404	180	472	550	110	141	175
Sept.	1,393	1,260	402	584	605	293	236	344
Oct.	1,304	1,034	550	268	280	345	66	250
Nov.	621	384	491	125	96	338	51	98
Dec.	306	25	71	70	18	180	43	17
Total quantity	8,078	7,329	4,441	3,299	3,159	2,173	2,024	901

Questions

1 Why was no excrement found in January and February?

2 In which environment do you think the most earthworms are found?

3 Is the amount of excrement always a direct indication of the number of earthworms present? Explain your answer.

4 Copy Figure 7-3 into your notebook. Plot the 12 values for the golf course and join the points with straight lines.

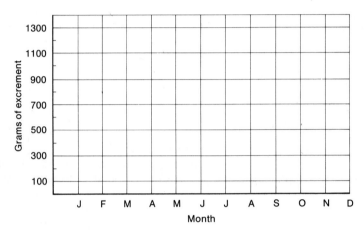

Fig. 7-3
Variations in earthworm excrement through a year.

5 Using other colors, do the same for the orchard and mixed forest environments. Account for any differences observed.

6 Why do the months of July and August show marked decreases in excrement?

7 How might the application of liquid manure have an effect on the worm population?

8 What factors might cause the very low amount of surface excrement noted in garden soils?

9 Account for the differences observed between the spruce forest and the mixed forest.

7.5 EFFECT OF CROPS ON WATER CONTENT IN THE SOIL

A study was performed to compare two fields, side by side. The only major difference was that one had been left fallow (unplanted in any crop), while the other was growing a lush crop of grain. After a week without rain, tests were carried out to determine the relative amounts of air, water, dirt, and stones. The percent composition was determined for particular depth ranges in the two fields. The graphs in Figure 7-4 summarize the data collected.

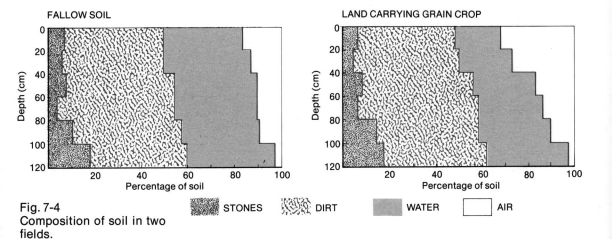

Fig. 7-4
Composition of soil in two fields.

STONES   DIRT   WATER   AIR

Questions

1 What is the major difference in the two fields with respect to water?

2 What does a comparison of air content indicate?

3 What caused the differences in the two fields?

4 What might you expect if you were to take a census of the soil inhabitants at each of the levels investigated in this survey? Why?

5 In a forest soil, would you expect the same results as shown in the field with the grain crop? Explain your answer.

6 Describe and account for the distribution of stones in the two fields.

7 Describe and account for the distribution of air in the two fields.

8 Do the distributions of stones and air appear to bear any relationship to the water content? Why?

7.6 DAILY CHANGES IN TEMPERATURE ABOVE AND IN SOIL

One of the important physical factors in the environment of any organism is the temperature. Most animals have daily fluctuations in activity that result from daily temperature changes, Human beings are no exception. When it gets too hot, the desire to work gets less. In some parts of the world the mid-day siesta is a daily break in the day's activity, designed to avoid unbearably hot temperatures.

Some humans are lucky in that they have the power to control their man-made environments, warming or cooling their places of work. Thus they maintain pleasant, workable conditions throughout the day. Some soil dwellers are equally lucky. Of course, soil organisms do not own air conditioners. But they

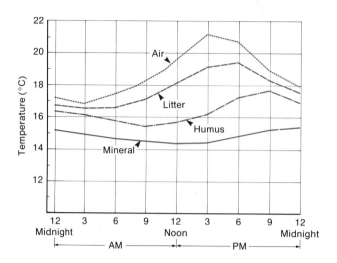

Fig. 7-5
Temperature readings taken over a 24-hour period at air, litter, humus, and mineral levels.

can make use of the physical characteristics of the soil to regulate their temperatures. The experiment described here provides information that should help you to discover how this is done.

An experiment was set up to measure temperatures in a mixed forest for a 24 hour period. Readings were taken at 3 hour intervals in air 30 cm above the ground, in litter at the surface, in the humus 5 cm below the surface, and in the mineral soil 15 cm deep. The results of this work are summarized in Figure 7-5.

Questions

1 At what time of day was the temperature highest in each of the four zones? Explain what the term "time lag" means by comparing the four sets of data in the graph. Account for the "time lags."

2 Determine what the "time lag" is for the litter on the surface and for the humus 5 cm below the surface. Estimate the "time lag" in the humus 2 cm below the surface.

3 Determine the approximate temperature range in the air, litter, humus, and mineral levels.

4 An animal that can survive large variations in the conditions of its environment is said to be *tolerant*. An organism that is *intolerant* cannot survive such changes. If an animal is intolerant with respect to temperature, it can live only within a narrow range of temperatures. In which zone, only, could a temperature intolerant animal live, if it were unable to move from one zone to the next?

5 Assume there is an organism, extremely temperature intolerant, which lives best at 17.5°C. If it is capable of active movement through the soil, where would it be at 12 midnight, 6 a.m., 12 noon, and 9 p.m. on the day of the experiment?

6 During a cloudless night, the air and soil surface quickly radiate back into space the heat that they absorbed during the day. Thus air temperatures often become very low at night, despite high daytime temperatures. Draw a graph similar to Figure 7-5 that you think would represent summer readings at a site in Death Valley, California. Plot only what you think would be the conditions in the air and in the soil at a depth of 15 cm. (Litter and humus layers are to be ignored. Why?) Account for the major differences between your graph and Figure 7-5.

7.7 TEMPERATURES: SHADING AND SEASONAL VARIATION

Many factors play a part in determining the temperature at any given point on the earth's surface. Elevation, latitude, and nearness to bodies of water are three that any geography student will be quick to mention. This study deals with two factors, each having specific importance to soil temperatures.

THE SHADE FACTOR

Temperatures were obtained in November in a fairly arid area of Nevada. At two different sites, temperature readings were taken at a number of heights above and below the soil surface. One site was shaded by juniper, while the other was not. The results are summarized in Table 6.

TABLE 6 TEMPERATURE READINGS IN AN ARID REGION OF NEVADA IN NOVEMBER

Condition	Height in cm from soil surface	Temperature in C°	
		Beneath forest cover	Unshaded field
Air	150	18	20
Air	90	18	21
Air	60	18	20
Air	30	18	21
Soil surface	0	16	33
Humus	−6	12	19
Mineral	−15	9	15
Mineral	−30	7	12

Questions

1 Construct a graph with height on the vertical (y) axis and temperature on the horizontal (x) axis. Plot the data and join the points with straight lines. Use different colors for the two sets of data.

2 At what time of day do you think the data were collected? What major difference would you expect 12 hours later?

3 From your own experience, which surfaces feel coolest and which hottest to the feet on a sunny day? Why?

4 The temperature data alone should indicate to you that there will be many differences in both the living and non-living features of the two sites. Predict what differences might be found if you were in the position to analyze the sites further. Give your reasons for each prediction.

THE SEASONAL FACTOR

Air is rather whimsical stuff. Its temperature fluctuates markedly on daily and yearly cycles. The surface soil layers show similar, but somewhat less extreme, changes in temperature. Soil inhabitants are fortunate that this is so, as you will gather from an analysis of the data in Table 7.

TABLE 7 AVERAGE MONTHLY TEMPERATURE ($°C$)
IN A MIXED FOREST IN MICHIGAN
FROM JULY TO FEBRUARY

	July	Aug.	Sept.	Oct.	Nov.	Dec.	Jan.	Feb.
Air	22.0	19.0	17.0	14.5	10.0	5.0	−1.0	−5.0
Litter	18.0	17.5	16.0	14.0	10.0	6.5	4.0	2.0
Humus	16.0	16.0	14.5	13.5	10.5	8.0	6.0	4.5

Questions

1 Construct a graph with temperature on the y-axis and the month on the x-axis. Plot the data for air, litter, and humus. Join consecutive points with straight lines, using a different color for each of the three regions.

2 Describe in words what the three sets of data indicate.

3 Snow started to accumulate at the study site in mid-January. How might a snow covering affect the temperature changes in the soil during the winter? Why?

4 Another site, close to the first one, remained windswept during the winter because it was more exposed. What differences will exist between the two sites as the winter progresses? What effects might this have on the soil inhabitants?

7.8 INVESTIGATION OF "LEARNING" IN ANTS

Normally ants must leave their nests to search for food. Of course, they must then find their way back across the surface to the nest. An ecologist, interested in behavior, wished to study how ants "learned" to navigate across the ground, since this is basic to the efficient operation of the ant society. He constructed a maze made up of two parts as shown in Figure 7-6. Each part is identical to the other only if the ants are allowed to enter from opposite ends. The B maze is entered from the side with the food, and the A maze is entered from the side with the nest.

Food

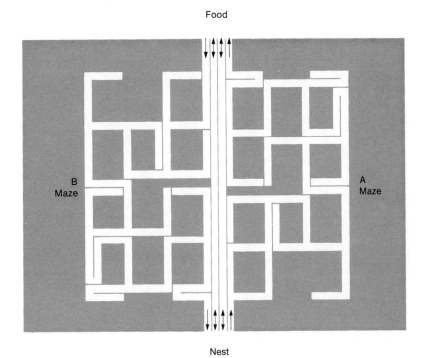

Fig. 7-6
An ant maze.

Nest

The experiment went as follows. Each of two ants was run through 30 trials. The first ant was allowed to go directly to the food but then had to return to the nest via maze B. Then 30 trials were conducted in which another ant had to first make its way through maze A to reach the food, and then was allowed to go directly to the nest, being prevented from returning by the maze. The results recorded were the number of errors made during each trial. An error occurred each time the ant was confronted with two possible routes and selected the wrong one.

An error also occurred if an ant turned around while following the correct route. The results for the two ants are given in Table 8.

TABLE 8

Trip no.	Errors in maze		Trip no.	Errors in maze	
	A	B		A	B
1	69	75	16	8	2
2	33	45	17	4	1
3	28	18	18	5	0
4	21	12	19	5	3
5	14	24	20	4	2
6	20	14	21	3	2
7	12	8	22	5	1
8	15	7	23	6	1
9	7	2	24	8	0
10	11	3	25	4	1
11	8	8	26	3	2
12	10	6	27	3	0
13	4	5	28	5	0
14	9	7	29	3	0
15	7	5	30	4	1

Questions

1 The maze looks rather complex, but actually there are not many points where a decision to go one way or the other has to be made. Look at maze B for 15 seconds. Then trace with your finger the shortest route to the nest. Take your finger away and count visually the number of places where you had to make a decision to go one of two possible ways.

2 Most people look at the maze and see fewer places to make a decision than there actually are. Can you explain why?

3 How is it possible for an ant to make as many errors as are shown in Table 8, if there are so few decisions to make? (Hint: Is it possible to make the same mistake twice? How?)

4 If a human were placed in a maze larger in size but identical in structure, how many mistakes would you estimate he might make the first time through? Can you compare the intelligence of the two organisms on this basis?

5 Construct a graph with the *y*-axis showing the number of errors on each trip and the *x*-axis, the trip number from 1 to 30. Plot the data for the two mazes, joining consecutive points with straight lines. Do the graphs reveal anything you did not already know?

6 Why is it important that the ant learn the route back to the nest better than the route to the food?

7 Could it be that ants purposely make "errors" going out to a food source? Is there any advantage in "exploratory" behavior?

8 Imagine the following. You are in a dense jungle. You can see less than 5 feet in any direction. You have only a can of paint. You know that within 50 feet of you there is a hole in the ground that leads to safety. You are afraid of moving in any direction for fear it will be the wrong way and you will end up farther away from the hole and get lost. What can you do to find the hole in perfect safety? How might this relate to ant behavior? What might ants possess as part of their anatomy?

9 If a second ant were placed in the maze after the first had "learned" the route, it would probably make less than 20 errors on its first trip. You might call this the "Hansel and Gretel trick." How does it work?

INDEX

Numbers in **boldface** represent pages with illustrations.

A

A horizons, 39–41, **39**, 44–47, **47**
Abiotic factors, 7, 18–21, 180
Absorption, water, **16**
Acidity, testing for, 113–114
    (*See also* pH)
Actinomycetes, 59, 93–94, **93**
Aeration, 29, 163, 164
Aerobic bacteria, 91
Aerobic respiration, 8, 9, 29
Algae, 3, 12, **15**, 20, **21**, 39, 59, 63, 83
Alkaline soils, 40
Aluminum, 43, 44, 46, 118, 119
Amino acids, 17, 18, **18**, 19
Ammonia, 18, **18**, **22**
Amoeba, 84, **84**, 89
Amphibians, 57
Anaerobic bacteria, 91
Anaerobic respiration, 8
Animals, interaction concept and, 3, 6, **6**
Annelids, 54–57
    See also Earthworms
Antibiotics, 94
Ants, 68, 71–73, **71**, 76
    Eciton hamatum, 174
    Formica rufa, 71
    Iridomyrmex humilis, 174
    "learning" in, case study, 191–193, **191**
    nests, 71–72, **72**, 174–175, 191
    reproduction, 71
    social behavior research, 174–175
    tests with, 140–148, **141**, **142**
Apatite, 115–116
Apterous (wingless) insects, 69–70, **69**, **70**
Arcella rhizopod, 84
Argentine ant (*Iridomyrmex humilis*), 174
"Armadillo trick," 67
Arthropods, 65–78, **66–78**, 132
    microscopic examination of, 145–148
    mounting on slides, 147–148
    numbers of, case study, 183–184, **184**

preservation of, 146
removing internal contents, 146–147
tests with, 137–148, **137**, **141**, **142**
    (*See also* names of arthropods)
Ascomycetes fungi, 87, 88
Ash, volcanic, 51
Ash tree, 182
Aspergillus fungus, 88, **88**
Auger, 99, **99**, 100, **100**
Automobiles, lead from, 171, 172
Autotrophic bacteria, 91–92
Autotrophs, 7, 8, 11, **11**
Azonal soil, **44**, 49, 51

B

B horizons, **39**, 41, 44, 45, **45**, 47, **47**
Bacilli cells, 90, **90**
Bacteria, 3, 9–11, 13. 16–19, **18**, 28, 39, 59, 63, 75, 89, **90**, **92**
    characteristics of, 90–91
    groups of, 91
    importance of, 90, 92–93
    reproduction, 91
    Rhizobium, 95
    Serratia marcescens, 83
    testing for, 153–159
Basidiomycetes fungi, 87, **87**
Bauxite, 46
Beech tree, 182
Beetles, 59, 68, 70, 74–76, **74–76**, 184
    importance of, 74–75
    larvae, 76
    nematode-eating, **9**, 10
Biological weathering, 34, 36, **36**
Biotic factors, 7, 10–14, 18, 180
Bodo flagellate, **85**
Bogs, 43, **50**, 127, 138
Bristletail, 70, **70**

C

C horizon, **39**, 41, 47, **47**
Caecilioides acicula snail, 64, **64**
Calcium, 36, 44, 50, 64, 114, 116, 117, 118
Calcium carbonate, 43, 46–47, 48
Calcium sulfate, 48
Capillarity, water, 15, **16**, 29, **29**, 30, **30**, **40**, 47, 50, 83
    test for, 106–107, **106**

Carabids (ground beetles), **74**, 75, 76
Carbohydrates, 7, 8, 11, 16, **17**
Carbon, 7, 15, **17**, 91, 112, 118, 177
Carbon cycle, 16–17, **17**
Carbon dioxide, 7, 8, 16, **17**, 19, 28, 31, 36, 40, 88, 115, 177
    tests for, 122–126, **123**, **124**
Carbonation, rock weathering by, 36
Carbonic acid, 35, 36, 40
Carnivores, 12, 13, 14
Carrion beetles (silphids), 75, **75**
Case studies: ants, "learning" in, 191–193, **191**
    arthropods, soil census of, 183–184, **184**
    decomposition of leaf litter, 180–183, **183**
    earthworm activity, 184–186, **186**
    temperature changes, 187–190, **188**
    water content, effect of crops on, 186–187, **187**
Cellulose, 73, 87, 89
Centipedes, 5, 14, **14**, 28, 65, 66, **66**, 68, 184
Cephalobus nematode, 58
Cercobodo flagellate, **85**
Chaetomium fungus, **86**, 87
Chemical weathering, 34, 35–36, 115
Chernozem soil, **44**, 47–48, **47–49**
Chestnut soil, **44**, 48, **48**, 49
Chilopoda, *see* Millipedes
Chipmunks, 79
Chlorophyll, 7
Cicada, 77, **77**
Ciliates, 84, 85, **85**
Classifications: azonal, **44**, 49, 51
    intrazonal, **44**, **49**, 50–51, **50**
    pedalfers, 43–47, **44**, **45**, **48**, **49**
    pedocals, 43, 44, **44**, 46–48, **47–49**
    zonal, 43–49, **44**, **48**, **49**, 51
Climate, interaction concept and, 6, **6**
Cocci cells, 90, **90**
Coleoptera, *see* Beetles
Colorimetric analysis, 112
Colpoda ciliate, **85**
Compost heap, 159–161
Consumer organisms, 11, 12, **14**
Core sampler. 99, **99**

194